Molecular Biotechnology

Edited by Sergey Sedykh

Published in London, United Kingdom

IntechOpen

Supporting open minds since 2005

Molecular Biotechnology
http://dx.doi.org/10.5772/intechopen.75339
Edited by Sergey Sedykh

Contributors
Na Wang, Galina Kovalenko, Larisa Perminova, Mitsuhiro Ueda, Kei Nakadoi, Shunsuke Sakurai, Kana Tsukamoto, Riddhi Prabha, Snehi Soy, Vinod Nigam, Stephen Rathinaraj Benjamin, Fabio De Lima, Yun Wang

First published in London, United Kingdom, 2021 by IntechOpen
IntechOpen is the global imprint of INTECHOPEN LIMITED, registered in England and Wales, registration number: 11086078, 5 Princes Gate Court, London, SW7 2QJ, United Kingdom
Printed in Croatia

British Library Cataloguing-in-Publication Data
A catalogue record for this book is available from the British Library

Additional hard and PDF copies can be obtained from orders@intechopen.com

Molecular Biotechnology
Edited by Sergey Sedykh
p. cm.
Print ISBN 978-1-83880-797-9
Online ISBN 978-1-83880-798-6
eBook (PDF) ISBN 978-1-83880-799-3

Meet the editor

Sergey Sedykh, Ph.D., graduated from the Faculty of Natural Sciences, Novosibirsk State University, in 2008. He defended his Ph.D. thesis in Biochemistry in 2012 at SB RAS Institute of Chemical Biology and Fundamental Medicine (Novosibirsk). He is a researcher at the Laboratory of Repair and Defense Systems of Novosibirsk State University, a researcher at Laboratory of Repair Enzymes of SB RAS Institute of Chemical Biology and Fundamental Medicine, and a leading specialist of educational projects and programs at Technopark of Novosibirsk Academgorodok. During 2013–2020, Dr. Sedykh managed projects supported by the Russian Science Foundation, Russian Foundation for Fundamental Research, President Council on Grants, and other research grant programs. His scientific interests include the biochemistry of human milk, catalytic antibodies, and exosomes.

Contents

Preface

Biotechnology is a unit of engineering biology that studies ways of using living organisms, their systems, or metabolic products to solve technological problems. Molecular biotechnology explores the possibility of creating living organisms with the necessary properties. Research in the field of molecular biotechnology has made it possible to improve traditional biotechnological production and the quality of food products, and has significantly increased the productivity of living organisms. More than 40 years ago, methods of genetic engineering in molecular biotechnology made it possible to use recombinant DNA and cell cultures in vitro.

In this book, the authors describe the modern trends and progress in biocatalysis and biotechnology. Three chapters cover the use of biocatalytic methods to solve the problems of processing renewable raw materials, the application of hydrolase-catalyzed reactions in organic synthesis, and the use of biocatalytic technologies in pharmaceuticals. The remaining three chapters describe electrochemical techniques for the detection of breast cancer biomarkers, the oxidation and photocatalytic treatment of textile wastewater, and the use of recombinant chitinase A, beta-GlcNAcase, and lytic polysaccharide monooxygenase for chitin hydrolysis. Overall, the book highlights the versatility of biotechnology and biocatalysis applications in science and industry.

In today's world, molecular biotechnology plays a leading role in processes of genetic engineering and gene editing, using genetically modified organisms in pharmacology, agriculture, biodegradation, and scientific research. The authors of the book address the application of molecular biotechnology in science and industry, diagnostics and therapy. This book will be of interest to researchers, students, and specialists in various fields of biological engineering.

Dr. Sergey Sedykh, Ph. D. in Biochemistry
Institute of Chemical Biology and Fundamental Medicine SB RAS,
Novosibirsk State University,
Russian Federation

Section 1

Biocatalysis

Chapter 1

Heterogeneous Biocatalysts for the Final Stages of Deep Processing of Renewable Resources into Valuable Products

Galina Kovalenko and Larisa Perminova

Abstract

Heterogeneous biocatalysis is a part of biotechnology and it has commercial potential for industrial implementation, in particular the final stages of deep processing of renewable raw materials. The commercially attractive heterogeneous biocatalysts are prepared by immobilizing practically valuable enzymatic active substances onto solid inorganic supports. Heterogeneous biocatalytic processes of the target conversion of substrate into valuable market product are carried out in periodic or continuous modes using traditional batch and packed-bed reactors, as well as novel types of vortex reactors in accordance with the principles of green chemistry. Heterogeneous biocatalysts for the final stages of deep processing of vegetable raw materials such as starch and oils are described here. One of the biocatalysts is glucoamylase immobilized by adsorption on mesoporous carbon support Sibunit™ type. This glucoamylase-active biocatalyst is used at the stage of starch saccharification, i.e., hydrolysis of dextrin to treacle and glucose syrups used in food and confectionary industries. The second of the biocatalysts is recombinant *T. lanuginosus* lipase immobilized on mesoporous silica KSK™ type and macroporous carbon aerogel. These lipase-active biocatalysts can effectively compete with traditional organic synthesis catalysts, and they are used in low-temperature processes carried out in unconventional anhydrous media such as interesterification of vegetable oils' triglycerides with ethyl acetate for producing ethyl esters of fatty acids (biodiesel and vitamin F) and esterification of fatty acids with aliphatic alcohols for synthesis of various esters used as fragrances, flavorings, odors, emollients, and nonionic surfactants in perfume and cosmetics industries. The prepared heterogeneous biocatalysts due to their high enzymatic activity and operational stability are promising for practical implementation.

Keywords: glucoamylase immobilized on mesoporous carbon, recombinant lipase immobilized on mesoporous silica and macroporous carbon aerogel, heterogeneous biocatalysts, starch saccharification, esterification of fatty acids

1. Introduction

Biocatalysis is a part of biotechnology that is inherently interdisciplinary and comprehensive, and its achievements are determined by the state of art in the fields of microbiology, molecular biology, biochemistry, chemical technology, and engineering sciences. Biocatalysis is a scientific field fully focused from the very beginning on practice, and the main task is to study and use only one selected enzymatic reaction for the purposeful target biotransformation of the initial reagent—substrate (S)—into the valuable product (P) demanded by the market [1]. Heterogeneous biocatalysis is a very important part of biocatalysis that is based on immobilized enzymatic active substances such as individual enzymes, whole microorganisms, and partially or completely disrupted microbial cells. Heterogeneous biocatalysis as an interdisciplinary sphere of professional activities, undoubtedly, has great scientific importance and commercial potential for industrial implementation, including the processing of renewable raw materials into valuable market products. In accordance with the 12 principles of green chemistry listed in [1, 2], biocatalytic processes satisfy all the requirements and provide environmentally friendly and energy-saving technologies that are a promising alternative to traditional chemical processes.

Despite unique catalytic properties of soluble enzymes, such as 100% selectivity and high reaction rates under very mild, usually ambient conditions, as well as chemo-, regio-, and stereo-specificity, their industrial applications are limited due to the main disadvantages, namely, homogeneous conditions of periodic processes and inability to reuse enzymes that fall into wastewater. Immobilization of enzymes on/in support may overcome this drawback. Immobilization is defined as the fixation of enzymatic active substances onto/inside water insoluble solid supports, accompanied by retaining their enzymatic activity at a high level and their significant stabilization (up to 10^3–10^5 times). Immobilization prevents the inactivation of enzymes and ensures their multifold reusability. The heterogeneous biocatalysts are prepared by immobilizing the enzymatic active substances. These biocatalysts are undoubtedly of great practical interest for widespread implementation in industrial periodic and continuous bioconversion processes using batch stirred-tank, packed-bed, and novel types of vortex reactors specially designed for the heterogeneous diffusion-controlled biocatalytic processes in order to overcome diffusion limitations and enhance biocatalysts' productivity.

It is generally recognized that heterogeneous biocatalytic processes are more commercially attractive for large-scale implementation than homogeneous technologies due to considerable simplification and reduction (in 1.2–1.4 times) of the total production cost. Of course, the cost of the final product decreases with a decrease in the cost of the enzymatic active component of the biocatalyst, as well as the support and the method of immobilization. In order to reduce all expenses, both not purified enzymes but partially or fully disrupted, or whole nongrowing microbial cells, as well as inorganic support such as silica and carbon, and adsorptive immobilization are preferable for preparing commercial biocatalysts.

For successful commercialization of the heterogeneous biocatalytic processes, for example, the process of starch dextrin hydrolysis (saccharification), a biocatalyst has to convert 45% of the substrate during 15–20 min with the inactivation half-time ($t_{1/2}$) of 30–120 days, which corresponds to 3–12 months operation of the reactor without reloading. A high value of $t_{1/2}$ (50–100 days) is essential to increase productivity up to the recommended value from 100 kg to 10 tons of final product per 1 kg of biocatalyst [3].

Heterogeneous biocatalysts for two important bioconversion processes are briefly described here. One of the biocatalysts is glucoamylase immobilized by adsorption on mesoporous carbon support Sibunit™ type. This glucoamylase-active biocatalyst is used at the stage of starch saccharification, i.e., hydrolysis of dextrin to treacle and glucose syrups. The second of the biocatalysts is recombinant *T. lanuginosus* lipase immobilized on mesoporous silica KSK™ type and macroporous carbon aerogel. These lipase-active biocatalysts can effectively compete with traditional organic synthesis catalysts, and they are used in low-temperature processes carried out in unconventional anhydrous media such as interesterification of vegetable oils' triglycerides with ethyl acetate for producing ethyl esters of fatty acids (biodiesel and vitamin F) and esterification of fatty acids with aliphatic alcohols for synthesis of various esters used as fragrances, odors, emollients, and nonionic surfactants in food, perfume, and cosmetics industries.

2. Glucoamylase-active heterogeneous biocatalysts for starch dextrin hydrolysis

The glucoamylases (glucan 1,4-alpha-glucosidases, EC 3.2.1.3) hydrolyze the glycosidic bonds at the end of polymer (starch) or oligomer (dextrin) chains releasing glucose. The main areas of industrial application of these enzymes are as follows: (1) a large scale two-step hydrolysis of raw starch successively catalyzed by α-amylase then glucoamylase, for production of sweeteners such as treacle and glucose syrups used in food industry, and (2) large-scale processes of hydrolytic conversion of starch to fermentable sugars as feedstocks for the production of some commodity chemicals and the first-generation biofuel such as bioethanol. These industrial processes are conducted on an enormous scale. Although the enzymes involved are relatively inexpensive, they are used on a single-use, throw-away basis. As mentioned above, immobilization of enzymes ensures enzyme recycling that can provide significant saving in the cost of final products not less than 20%.

The glucoamylases are the main enzymes used in the key second stage of starch conversion—hydrolysis of dextrin (saccharification), following the stage of liquefication (dextrinization) of starch by amylases. It should be noted that the immobilization of glucoamylase (not amylase) is justified and appropriate because this enzyme converts relatively low molecular weight substrates such as dextrin (3–5 kDa), and diffusion limitations can be overcome, in particular via design of employed reactor. The development of heterogeneous biocatalysts with high glucoamylase activity and operational stability is of great importance since they can serve as the basis for modern technology for deep processing of renewable vegetable raw materials into demandable sweeteners.

Back in 1970s, Corning Glass Co. carried out the fist pilot plant tests of a packed-bed reactor filled with a heterogeneous biocatalyst prepared by covalent immobilization of glucoamylase on macroporous silica; the glucose productivity was 450 kg/day. The main requirement for the commercial glucoamylase-active biocatalysts was a sufficiently high operational stability at pasteurization temperature of 60°C or higher. Inactivation half-times ($t_{1/2}$) of the tested Corning Glass biocatalyst were 520, 150, and 75 h at 55, 60, and 70°C, respectively [4]. The best result described later in 2000 for a biocatalyst prepared by immobilization of glucoamylase on polystyrene was that $t_{1/2}$ = 300 h at 50°C [5]. The best result described in the recent papers during 2008–2019 is that the glucoamylase immobilized by formation of cross-linked enzyme aggregates (CLEA) has "excellent recyclability, retaining over 45% of the relative activity after 24 runs" over a broad range of temperature (55–75°C) [6]. According to the opinion of the specialists working on the R&D projects of heterogeneous stage of saccharification the low thermal stability of the immobilized glucoamylase at elevated temperature (above 50°C) was the main reason why this process has not been commercialized so far.

Reputedly, the inexpensive and available carbonaceous materials with appropriate texture parameters are promising supports for adsorptive immobilization of enzymes, in particular of glucoamylase, for the preparing of commercially attractive heterogeneous biocatalysts. Sibunit™-type supports from a new class of carbonaceous materials are porous carbon-carbon composites that combine the advantages of both graphite such as chemical stability and electrical conductivity, and active carbons in particular high specific surface area and adsorption capacity. These supports are characterized by a high volume of mesopores and a narrow controllable pore size distribution; some types have a large proportion of pores with a size of 5–20 nm or a bidisperse meso- and macroporous structure suitable for enzyme immobilization. Indeed, the sizes of globular molecules of most enzymes in aqueous solutions are approximately 10 nm. And it can be argued that the specific accessible surface area can be calculated using pore size distribution diagrams on the assumption that pores with a diameter more than 10–15 nm are available for immobilization of enzymes.

Textural parameters of the Sibunit support are as follows: total specific surface area, $S_{sp\ BET}$ = 550 m^2/g; total pore volume, V_Σ = 0.86 mL/g; and average pore diameter, D_{pore} = 18 nm as follows from pore size distribution diagram (**Figure 1a**). Specific surface area accessible for enzyme immobilization was estimated to be 92 m^2/g, that is 16% of $S_{sp\ BET}$. Surface of Sibunit formed by round coke deposits of pyrolytic carbon looks rough and porous on the scanning electron microscopic (SEM) images (**Figure 1b**).

The properties of the best glucoamylase-active biocatalyst (designated as GlucoSib) prepared by physical adsorption of commercial enzyme preparation GlucoLux™ type (produced by Sibbiopharm, Novosibirsk, Russia) on mesoporous carbon support Sibunit™ are described in [7] and briefly here.

It was found that the porous texture of the carbon supports plays a key role in stabilization of glucoamylase [7]. Predominant mesopores (10–20 nm in diameter) of Sibunit and bulk catalytic filamentous carbon are appropriate in size to hydrated enzyme molecules, whereas micropores of activated carbon (4 nm in diameter) are too small. Exactly, mesopores provide multipoint binding enzyme molecules inside the support and, as a result, ensure great stabilization of the activity. The thermal stability of glucoamylase adsorbed on mesoporous carbon supports, measured in dextrin solutions (32–53 w/v%) was found to be ~10^5 times higher than for soluble enzyme [7].

Macrokinetics of dextrin hydrolysis by immobilized glucoamylase was investigated. Internal diffusion of dextrin inside the porous space of the biocatalysts toward adsorbed glucoamylase was found to be a rate-limiting stage of the saccharification process. Indeed, the rate of dextrin hydrolysis significantly reduced if the granule size was larger than 1 mm; for example, from 400 to 180 U/g for 1.2- and 3-mm granules, respectively [7]. The GlucoSib-type biocatalysts were prepared using support granules with diameter less than 1 mm; for example, the activity of the

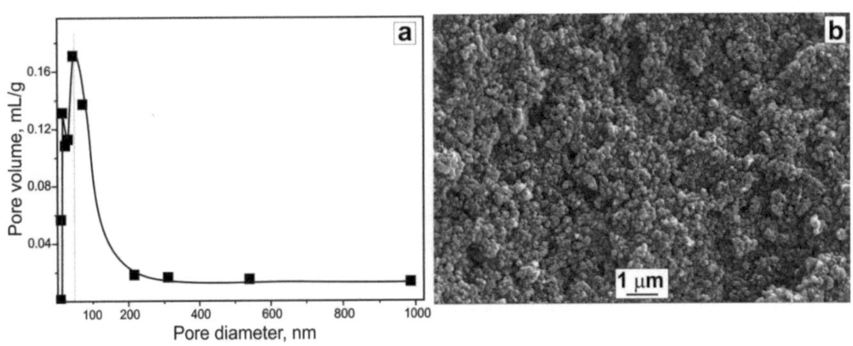

Figure 1.
(a) Pore size distribution diagram and (b) SEM image of Sibunit surface.

biocatalyst prepared by adsorption of glucoamylase on Sibunit granules of 0.2–0.7 mm in size was determined to be maximal, 530 U/g.

In order to overcome external diffusion limitations, a novel type of reactor such as immersed vortex reactor (IVR) was designed and tested in lab scale [7, 8]. The reactor body was filled with biocatalyst granules and then was immersed in a substrate solution and rotated. Substrate solution of dextrin was sucked through a bottom hole upon the body rotation and then moved with various high speeds through the biocatalyst bed toward side holes. Thus, a significant intensifying mass transfer of a substrate toward immobilized enzymes and, as a result, overcoming diffusion limitations was achieved by rotating the reactor body immersed in a substrate solution. Another very important advantage of the vortex type of reactor is an absence of stagnant zones and jet streams inside the bed of biocatalyst. To prevent formation of stagnant zones, the profiled reactor body was designed (**Figure 2a**). Since the profile of each half of the prefabricated reactor body was made in accordance with hyperbolic equation, the annular cross-section area for a substrate solution stream was equal to $2\pi R \cdot h$ and remained constant during circulation of substrate solution under operation. During this movement, the liquid was obviously affected by hydrodynamic, centrifugal, and inertial (Coriolis) forces, which resulted in a vortex flow of the substrate solution. The centrifugal forces were responsible for slight compression of the biocatalyst bed, narrowing channels for liquid flowing between granules, which results in the additional increase of mass transfer. Because of high mechanical strength of the Sibunit support, the granules were not distorted during the operation. Thus, the formation of stagnant zones was minimized.

A lab scale setup of the immersed vortex reactor filled by the glucoamylase-active biocatalysts GlucoSib was studied in the heterogeneous process of starch dextrin hydrolysis [7]. To elucidate optimal conditions for the IVR operation, the effect of rotation rate of the IVR body on the activity was studied. The maximal activity of GlucoSib, 700–750 U/g, was measured at body rotation of 300–900 rpm (**Figure 2b**). At 1000–1200 rpm, the reaction rate decreased perhaps due to the formation of a funnel in the rotating reaction medium, and special profiled device was designed to remove this defect.

The stability of the glucoamylase-active biocatalysts determined during continuous operation in saccharification heterogeneous stage was sufficiently high. Thus, the half-life time ($t_{1/2}$) exceeded 700 and 350 h at 50 and 60°C, respectively.

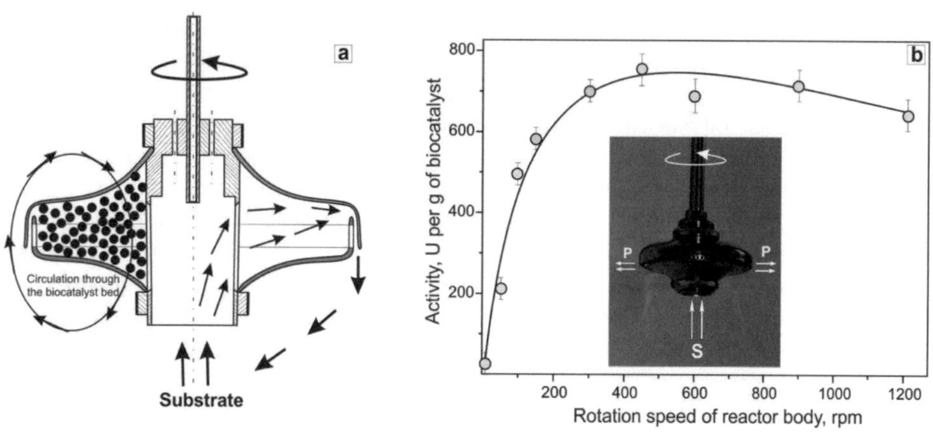

Figure 2.
Lengthwise cut of IVR body (a) and glucoamylase activity of the biocatalyst depending on rotation rate of reactor body (b). Photo of IVR body. Conditions of dextrin hydrolysis: 50°C, 0.05 M acetate buffer pH 4.6, 10 w/v% solution of potato dextrin as substrate.

Comparing these data with the Corning's results in [4], the $t_{1/2}$ of GlucoSib biocatalysts was estimated to be higher, 350 h vs. 150 h at 60°C, respectively. Long-term stability was sufficient also; the biocatalysts retained initial activity for 10 months upon storage at ambient temperature (18–22°C).

A technological scheme has been proposed and tested on a laboratory scale using the GlucoSib biocatalyst and immersed vortex reactor IVR for production of starch treacle and glucose syrups by heterogeneous dextrin hydrolysis. The advantages of this technological scheme are as follows: (1) significant acceleration of dextrin hydrolysis; (2) energy and resource saving in comparison with traditional starch processing; (3) a high quality of the final products due to the lack of protein impurities; and also, very importantly, (4) easily regulated carbohydrate composition of the treacle by simply stopping rotation of reactor body. It should be noted that when comparing the efficiency of the process of dextrin hydrolysis in vortex reactor with parameters of traditional packed-bed reactor the productivity of the IVR was higher by 1.2–1.5 times. The productivity in a novel proposed technology was calculated to be 5.3 tons of glucose per 1 kg of biocatalysts GlucoSib that is quite commercially attractive.

2.1 Conclusion for the part 2

The highly active and stable heterogeneous biocatalysts for dextrin hydrolysis were prepared by adsorption of glucoamylase on mesoporous carbon support Sibunit™. Under technological conditions (32 w/v% dextrin, 60°C, pH 5), the maximal activity was observed to be equal to 750 U/g, and inactivation half-life time ($t_{1/2}$) was 350 h. The immersed vortex reactor designed specially for the biocatalytic diffusion-controlled heterogeneous processes was used to carry out starch saccharification with enhanced productivity roughly estimated as 5 tons of glucose per 1 kg of biocatalysts.

3. Lipase-active heterogeneous biocatalysts for vegetable oil and fatty acid bioconversion

Heterogeneous lipase-active biocatalysts are of great importance due to remarkable properties of their enzymatic active component, namely lipase (triacylglycerol acyl hydrolase, EC 3.1.1.3.), that catalyzes a variety of reactions involving triglycerides—hydrolysis, esterification (synthesis of esters), inter-(or trans-)esterification (intra- and intermolecular exchange between fatty acids' residues), alcoholysis, and aminolysis [2, 9]. Nowadays, the main areas of industrial application of lipases and lipase-active biocatalysts are as follows: (1) hydrolysis of oily and fat stains on clothes by smart washing powders containing lipases as additives; (2) a large scale process of interesterification of the oil-fat blend to produce valuable ingredients, in particular for various spreads and margarine; (3) alcoholysis (methanolysis/ethanolysis) of triglycerides of vegetable rapeseed and soybean oils, as well as waste cooking and algal oils, into methyl/ethyl esters of fatty acids for production of biodiesel as an additive to fuel for engines; and (4) esterification of organic acids and synthesis of marketable valuable products including various esters of fatty acids for food and cosmetic industries, as well as enantiomers for pharmaceutical industry.

The most interesting and unique property of lipases is their ability to catalyze reaction in anhydrous media of organic solvents with water content less than 1%. Despite nonconventional conditions, these processes are characterized by high efficiency such as 100% selectivity, high conversion, and yield of final product.

One of the best results of oils' methanolysis by commercial biocatalyst Novozym® is that conversion of triglycerides into methyl esters of fatty acids (biodiesel) was ~99% for 50 h at 45°C [10]. In 2007, the first biocatalytic process of the biodiesel production was implemented by methanolysis of edible oils' waste with the productivity of 10,000 tons; the biocatalyst was prepared by immobilization of lipase from *Candida* sp. [11].

NOVO (NOVOZYMES) Company is a leader in the production and sale of heterogeneous biocatalysts prepared by immobilizing the recombinant lipases on various supports. The Lipozyme® TL IM biocatalyst is prepared by immobilization of recombinant thermostable 1,3-specific *T. lanuginosus* lipase on silica. This biocatalyst is widely used in the industrial interesterification of fat-oil blends in order to produce valuable products, such as specialized fats and spreads without undesirable *trans*-isomers of fatty acids, as well as substitutes of cocoa butter and dairy fat. The Novozym®435 biocatalyst is prepared by immobilization of nonspecific *Candida antarctica* lipase on a macroporous polyacrylic polymer. The commercial Novozym® biocatalysts type are used in biodiesel production by methanolysis of vegetable oil (rapeseed and soya been) and waste oils of cooking. Nowadays, the NOVO biocatalysts are intensively studied for application in various processes, including organic synthesis.

It is well-known that methanol and ethanol inactivate enzymes rapidly. Therefore, in order to reduce the biocatalysts' inactivation, methanol was added stepwise in small portions during the reaction cycle of the biodiesel production. Another acylating reagent—methyl or ethyl acetate—was examined to be used. It was found that this reagent did not reduce the activity of the commercial biocatalyst Novozym® even at a molar ratio of oil to methyl acetate equal to 1:12; and under optimal operating conditions, the yield of methyl esters was equal to 96% and the biocatalyst's half-life time ($t_{1/2}$) increased 20-fold in comparison with $t_{1/2}$ in reaction with methanol [12]. Acyl derivatives of glycerol produced in interesterification of vegetable oils with methyl or ethyl acetate are valuable commercial products also. For example, mono- and triacyl glycerol are employed as fuel additives. Triacylglycerol (triacetin) is widely used in the food industry due to its good moisture-retaining properties. If the linseed oil is used in biocatalytic interesterification with ethyl acetate, the produced mixture of ethyl esters of ω3-, ω6-unsaturated fatty acids (vitamin F) is a valuable product for cosmetics industry and fodder additives production.

Nowadays, enzymatic esterification is considered as a competitive alternative to the chemical organic synthesis of various esters that are valuable commercial products commonly used in manufacturing flavors, fragrances, emollients, lubricants, antimicrobial agents, and nontoxic surfactants. The requirement of consumers for such natural products is constantly increasing. Compared with organic synthesis using strong liquid and solid acids as catalysts and temperature above 100°C (usually, 120–150°C), the enzymatic esterification is currently of great commercial interest since this method of esters' production proceeds efficiently at a low temperature (usually, at 20–40°C) without the formation of any by-products and with high specificity toward substrates. The heterogeneous lipase-active biocatalysts for the low-temperature esters' synthesis are prepared, as mentioned above, by immobilizing lipases on solid supports by various chemical origins and texture. These heterogeneous biocatalytic processes realized in periodic or continuous modes in anhydrous reaction media fully satisfy the requirements of "green" chemistry and they are promising for implementation into the organic synthesis industry [2].

The authors and their collaborators have developed and researched systematically the heterogeneous biocatalysts in which the enzymatic active component was a recombinant *T. lanuginosus* lipase. These lipase-active biocatalysts were prepared both by entrapment of fully disrupted cells (lysates) of recombinant strain-producer r*E.coli*/lip inside silica xerogel and its nanocarbon-containing composites [13, 14, 17] and by adsorption of *T. lanuginosus* lipase produced by recombinant lipase on silica [15–18] and carbon aerogel [19]. These biocatalysts were studied in the reactions of tributyrin hydrolysis [13, 19], interesterification of oil-fat blends and vegetable oil triglycerides with ethyl acetate [13, 14], and esterification of fatty acids [15–19]. The two biocatalysts prepared by adsorptive immobilization of recombinant *T. lanuginosus* lipase are described briefly below.

3.1 Biocatalysts prepared by adsorption of recombinant lipase onto mesoporous silica

The enzymatic active component of the developed heterogeneous lipase-active biocatalysts was recombinant *T. lanuginosus* lipase (designated as r*Pichia*/lip) that was produced extracellular by the methylotrophic yeast *Pichia pastoris* X-33 strain specially constructed by the following genetic engineering manipulations: (1) chemical synthesis of the gene of a mature *T. lanuginosus* lipase taking into account the nucleotide sequence found in the Protein Data Bank (PDB-database); (2) cloning of the synthesized gene into a plasmid vector and production of the constructed recombinant plasmid in *E. coli* cells; (3) transformation of competent *P. pastoris* cells with the obtained plasmid and selection of recombinant yeast clones; and (4) analysis of the selected clones for the ability to produce extracellular and secrete recombinant *T. lanuginosus* lipase into the nutrient culture medium. Finally, conditions for the cultivation and intensive growth of the r*Pichia*/lip strain-producer were optimized in order to increase the concentration of the secreted target enzyme. The lab scale production of recombinant lipase was carried out in a 10-liter BIOK gas-vortex bioreactor (ZAO Sayany, Russia); the lipase concentration in nutrient media reached 2 g/L. Partial purification of recombinant lipase was carried out by precipitation of the secreted r*Pichia*/lip with ammonium sulfate (up to 75% saturation) at 4°C for 16 h. The precipitates were dissolved in distilled water, and further dialysis against a 25 mM acetate buffer pH 4.0 was carried out. The samples of dialyzed and lyophilized r*Pichia*/lip were used for the preparing lipase-active biocatalysts.

Mesoporous silica (SiO_2) of KSK™ type was used as a support for immobilization of recombinant *T. lanuginosus* lipase; its textural parameters are as follows: the specific surface area 157 m^2/g; total pore volumes (V_Σ) 0.76 mL/g; average pore diameters 19 nm; and porosity 58%. The lipase-active biocatalysts were prepared by both "spontaneous" and "forcible" adsorption using r*Pichia*/lip solutions at the same concentration. Spontaneous adsorption was carried out while contacting support with lipase solution at a ratio of support weight to solution volume of 1:(3–10) for 24–48 h. For such a long contact, adsorptive immobilization of lipase occurred due to electrostatic and acid-base interactions of enzyme molecules with silica surface. Forcible adsorption of lipase was carried out via moisture capacity impregnation of mesoporous silica by lipase solution, followed by drying granules. When biocatalyst was dried, enzyme molecules lost their hydrated shells and stuck to the surface. In both cases, the amounts of adsorbed enzyme were close in magnitude and equal to 4.2 and 5.4 mg/g for spontaneous and forcible adsorption, respectively. It was found that under exactly the same reaction conditions of esterification and synthesis of *n*-butyl heptanoate, the stationary activity of the spontaneously adsorbed lipase was two-fold less than the activity of forcibly adsorbed enzyme, for example, 2.4 vs. 5.0 U/g, respectively (recall that 1 U = 1 μmol/min). The method of forcible adsorption was characterized not only by comparatively high enzymatic activity but also by simplicity of its implementation and economical enzyme consumption. For example, a minimal volume of lipase solution used for spontaneous adsorption on 1 g of silica was 3.0 mL, whereas for forcible adsorption, it was 0.8 mL equal to total pore volume (V_Σ) of SiO_2. All results described here referred to the lipase-active heterogeneous biocatalysts (designated as LipoSil) prepared by forcible adsorption on silica of recombinant lipase, which were used predominantly in esterification processes [15, 16, 18].

Biocatalytic processes of enzymatic esterification were performed at ambient conditions (20 ± 2°C, 1 bar) in unconventional anhydrous media of organic solvents such as hexane and diethyl ether. The saturated fatty acids differing in the number of carbon atoms (C2–C10, C18), as well as aliphatic alcohols differing in the structure of the molecules, namely, the number of carbon atoms (C2–12, C16), the isomerism of the carbon skeleton (*n-* and *iso-*) and OH-group position (*prim-*, *sec-*, and *tert-*) were studied as substrates in esterification by LipoSil. There were some peculiarities of operation of the preliminary dried lipase-active biocatalysts in nonaqueous organic solvents. A considerable increase of the activity was observed during the 1st–3rd reaction cycles. This phenomena, named preconditioning stage, was due to the ongoing accumulation of formed product—water—in the vicinity of the adsorbed lipase inside the silica-based biocatalyst, and this stage proceeded faster and the higher than the activity of the biocatalyst. For example, if the biocatalytic activities were about 5 and 500 U/g, then the activation of the biocatalysts (preconditioning) proceeded within 24 and 0.5 h, respectively. Calculation showed that under studied conditions upon full conversion of fatty acid, maximal 0.1 mL of water was formed inside for one reaction cycle. Since the total pore volume of silica (0.8 mL/g) was multifold greater than the volume of the water formed, this amount of H_2O was firmly held inside KSK™ silica commonly applied as a dehumidifier for industrial gases. Therefore, during esterification, the favorable aqua microenvironment was created for adsorbed lipase, and the activities of the dried biocatalysts increased by 2–4 times. After preconditioning stage, the biocatalytic activity, named stationary, was measured in batch reactor during several tens of reaction cycles. Each reaction cycle was completed to the full conversion of acid, close to 85–90%. Then, the reaction medium was removed by decantation and the biocatalysts were washed by solvent for 20 h. The next reaction cycle was started by adding fresh reaction medium containing substrates of lipase—acid S_1 and double molar excess of alcohol S_2. Stationary activity of the biocatalysts was fluctuated in magnitude during the consecutive reaction cycles of the periodic esterification process, perhaps due to the presence of ester (product) residues inside the operating biocatalyst. As it can be seen in **Figure 3**, the operational stability of the prepared biocatalyst was sufficiently high; its stationary activity was retained completely after 38 cycles (~900 h) of esterification of various fatty acid. Also, the biocatalysts possessed a high long-term stability; the activity was determined to be ~80% of initial one after storage for 9 months in the solvent (hexane and diethyl ether) at ambient temperature and in dried state in refrigerator.

Obviously, a very important property of heterogeneous biocatalysts is the high operational stability, since the productivity calculated by multiplying average activity by $2 \cdot t_{1/2}$ increases significantly. The esterifying activity of the prepared lipase-active biocatalysts did not practically change during 500–1000 h of operation. Under studded conditions, the productivity was evaluated as ~2 tons of product per 1 kg of LipoSil biocatalyst.

The study of the functional properties of enzymes after their immobilization, such as activity, stability, and, importantly, specificity, is of great interest. Of particular scientific and practical interest is the research of the possibility of modulating these properties, by engineering heterogeneous biocatalysts, in particular, by

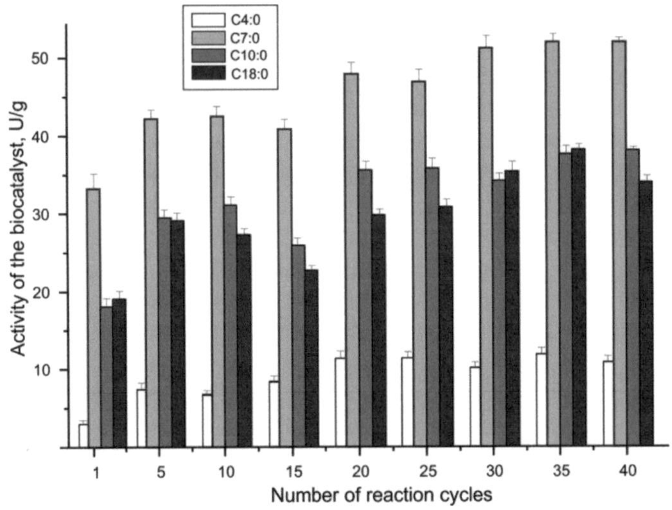

Figure 3.
Operational stability of the lipase-active biocatalyst in periodic esterification of various fatty acids such as butyric (C4:0), enanthic (C7:0), capric (C10:0), and stearic (C18:0) with n-butanol as a function of enzymatic activities depending on a number of reaction cycle. Conditions of esterification: 20 ± 2°C, 0.25 M acid, 0.50 M alcohol, hexane:diethyl ether = 1:1, content of biocatalyst in reaction medium 20.8 wt.%. The amount of adsorbed lipase is 14.0 mg/g.

Acids ↓	Alcohols →								
	C2	C3	C4	C5	C8	C10	C11	C12	C16
C2	0.00	–	–	0.03	0.02	–	–	–	0.02
C4	0.08	0.21	0.21	0.19	0.15	0.19	0.08	0.06	0.04
C5	0.09	0.23	0.28	0.15	0.17	0.28	0.10	0.08	0.07
C6	0.08	0.26	0.31	0.23	0.19	0.47	0.09	0.07	0.07
C7	0.54	0.98	1.00	0.88	0.85	0.81	0.72	0.53	0.47
C9	0.27	0.64	0.54	0.52	0.56	0.59	0.28	0.24	0.19
C10	0.30	0.57	0.64	0.63	0.58	0.54	0.32	0.26	0.27
C18	0.39	0.57	0.78	0.61	0.57	–	0.39	0.34	0.39

Table 1.
Matrix of relative activities of the immobilized recombinant T. lanuginosus lipase in reaction of esterification of various pairs of substrates—saturated fatty acids and aliphatic primary n-alcohols.

selecting the chemical nature of the supports [20, 21]. In our research, the specificity of heterogeneous enzymatic esterification was determined by comparing the reaction rates for various pairs of substrates—saturated fatty acids and aliphatic primary *n*-alcohols—and the matrix of relative units of activities for the LipoSil biocatalyst was composed (**Table 1**) [18].

The following conclusions can be drawn from the data presented in **Table 1**: (1) the rate of esterification of acetic (C2) acid is very low; thus, it is practically impossible to obtain acetate esters; (2) the rate of synthesis of ethyl ester (reaction with ethanol C2) is 1.3–2 times lower than that in reactions with C3–C16 alcohols; (3) the esterification of fatty acids with a number of carbon atoms more than six (C6) occurs 2–5 times faster than that of low molecular weight fatty acids C4–C6; and (4) the maximal observed rate is determined for esterification of enanthic (C7) acid with C3–C8 alcohols. Also, one can see that the immobilized on silica recombinant lipase is more sensitive to the molecular structure of saturated fatty acids than

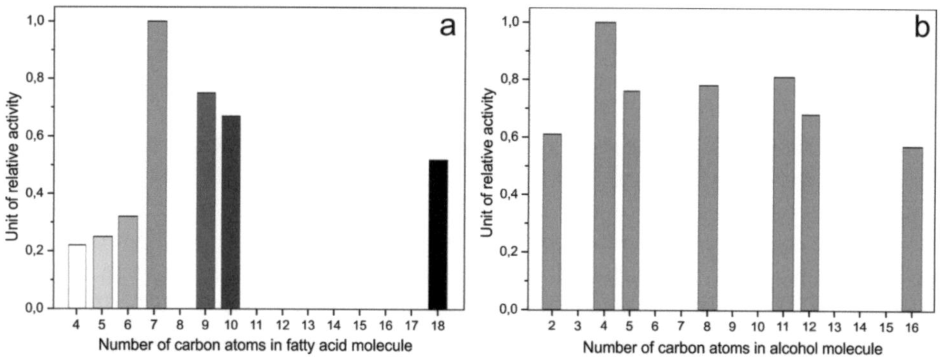

Figure 4.
Relative activities of the immobilized recombinant T. lanuginosus lipase in reaction of esterification of (a) different saturated fatty acids with n-butanol, and (b) enanthic acid with different alcohols. Conditions of esterification: 20 ± 2°C, 0.25 M acid, 0.50 M alcohol, hexane:diethyl ether = 1:1, content of biocatalyst in reaction medium 20.8 wt.% the amount of adsorbed lipase 14.0 mg/g.

that of aliphatic alcohols because the maximal differences in rates were ~6 times for various acids (**Figure 4a**) and only ~2 times for alcohols (**Figure 4b**).

It was found that immobilized on silica r*Pichia*/lip was not sensitive to isomerism of alcohols' molecule, and the rates of esterification of fatty acids with primary *n*-C4,C5 alcohols were almost 10 ± 4% lower than those with *is0*-C4,C5 alcohols [18]. The most pronounced specificity was observed toward the position of OH-group in alcohol molecules. The rates of esterification of fatty acids with secondary alcohol (*sec*-propanol and *sec*-butanol) was two order of magnitude lower than those with primary alcohols, for example, 0.36 U/g vs. 56.1 U/g for *sec*- and *prim*-propanol, respectively [18]. The esterification of fatty acid with ternary alcohols (*tert*-butanol) did not occur, and reaction rate was zero. Also, it was found that the substrates (both acid and alcohol) with aromatic and cyclic restudies reacted with very low rates [15, 19].

The kinetic parameters such as Michaelis constant (K_M) for acid and maximal reaction rate (V_{max}) under studied conditions were determined in esterification of enanthic (heptanoic, C7:0) acid with a double excess of *n*-butanol. The kinetic curve as a function of the initial reaction rate from the initial fatty acid concentration lower than 1.0 M was satisfactory approximated by the classic hyperbolic Michaelis equation, $V = \frac{V_{max} \cdot C_0}{K_M + C_0}$, where V and V_{max} are the initial observed and maximum reaction rates, respectively, μmol/L·s^{-1}; C_0 is the initial substrate concentration, mol/L; and K_M is the Michaelis constant, mol/L. At a concentration of enanthic acid above 1.0 M, the reaction rate decreased perhaps due to inactivation of enzyme by high concentration of *n*-butanol as described previously for esterification of capric acid with *iso*-pentanol [17]. The values of K_M and V_{max} determined using Lineweaver-Burk linear approximation and the regression of hyperbolic equation by soft Origin' programs were 0.22 ± 0.05 mol/L and 66.7 ± 4.0 μmol/L s^{-1}, respectively. A comparison with data published earlier in [16, 17] showed that the Michaelis constant for enanthic (C7) acid was ~3 times less than K_M for capric (C10) acid during esterification with C4,C5 aliphatic alcohols. This means, that the affinity of adsorbed recombinant lipase toward enanthic acid is higher; thus, the maximum rate V_{max} will be achieved at lower initial fatty acid concentrations, and it is important for practice.

3.2 Conclusion for the part 3.1

The catalytic properties of the lipase-active heterogeneous biocatalysts such as enzymatic activity, stability, and substrates specificity were investigated in the esterification of various saturated fatty acids with aliphatic alcohols. These biocatalysts were prepared by immobilizing recombinant *T. lanuginosus* lipase onto mesoporous silica by impregnation method. Particular attention was paid to the study of the substrate specificity of immobilized lipase. It was found that forcibly adsorbed on silica *T. lanuginosus* lipase demonstrated broad substrate specificity. Saturated fatty acids with a number of carbon atoms 6 or more (till 18) and alcohols with a number of carbon atoms 3 or more (till 16) reacted with comparatively high reaction rates. The reaction rates depended slightly on isomerism (*n*- and *iso*-) of carbon skeleton of C4-C5 alcohols, whereas the rates depended strongly on the position of OH-group, and secondary and ternary alcohols did not react with fatty acids. Comparing the rates of esterification of various pairs of substrates using primary alcohols, a matrix of relative biocatalytic activities was composed. According to this matrix, the rates of synthesis of C4-C18 esters were sufficiently high, while the rate of synthesis of acetate ester was very low. Under the same reaction conditions, the maximal rate was observed in esterification of enanthic (C7:0) acid with butanol. The classical Michaelis-Menten kinetics was inherent for biocatalytic esterification of this acid with double molar excess of alcohol, and the main kinetic parameter, Michaelis constant for acid, was determined to be 0.22.mol/L.

It is well known that, from a practical point of view, the stability of heterogeneous biocatalysts is a very important characteristic that determines the biocatalyst productivity. The prepared lipase-active biocatalysts possessed considerably high operational stability in a periodic batch process of low-temperature esters' synthesis carried out in unconventional anhydrous media of organic solvents (hexane and diethyl ether). The enzymatic activity of the biocatalysts was completely retained for several tens of reaction cycles.

It was concluded that due to the remarkable catalytic properties combined with simplicity of immobilization method, the prepared lipase-active biocatalysts are promising for practical use in organic synthesis including the production of valuable esters.

3.3 Biocatalysts prepared by adsorption of lipase on carbon aerogel

Immobilization of recombinant *T. lanuginosus* lipase (SIGMA Co.) was carried out by spontaneous adsorption onto macroporous carbon aerogel [19]. It should be noted that carbon aerogels are a unique class of porous materials with a very low density (less than 0.1 g/mL) and a porosity of up to 90–99%. These materials are novel promising adsorbents for enzyme immobilization with great potential for practical implementation. Macroporous carbon aerogel (MCA) for our research has been produced by *in situ* synthesis of multi-walled carbon nanotubes (CNTs) via catalytic high-temperature decomposition of ethylene over the supported Fe:Co catalyst. The carbon aerogel was obtained in the form of ball-shaped granules of 1–10 mm in diameter (**Figure 5a**).

The three-dimensional framework and rigid macrostructure of MCA were formed by chaotic interlacing carbon nanotubes (**Figure 6a**). The number of walls and the diameter of CNTs determined by high resolution transmission electron microscopy (HRTEM) were equal to 12–14 and 15–25 nm respectively (**Figure 6b**). The density of MCA was 0.06 g/mL. The texture parameters of carbon aerogel were as follows: the specific surface area (S_{BET}) was 80–110 m^2/g, total pore volume (V_{Σ}) was 10-14 mL/g; macropores of 0.5–1 μm in diameter were predominant in texture; and the volumes of meso- and micropores did not exceed 2% of V_{Σ}.

When studying adsorption of recombinant lipase on MCA, it was found that adsorption graph contained two "plateaus" corresponding to adsorption of ~100 mg/g, or ~0.8 mg/m^2 of carbon aerogel for the 1st plateau, and 200 mg/g for

Figure 5.
(a) Photo of granules of macroporous carbon aerogel and (b) photo of agglomeration of fine dispersed carbon nanotubes (CNTs) under the influence of increasing concentration of tributyrin.

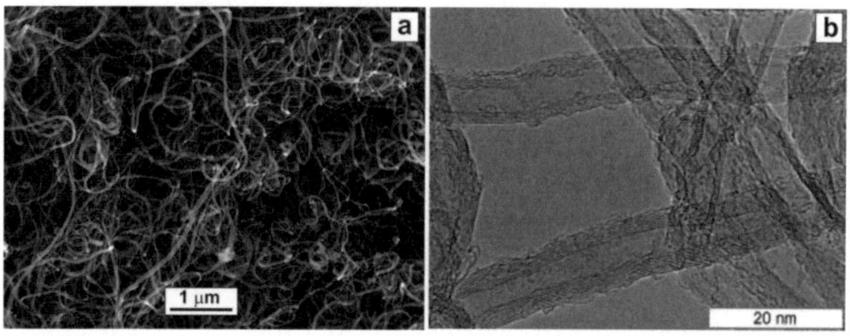

Figure 6.
(a) SEM image of the inside of the granules of macroporous carbon aerogel; (b) HRTEM image of carbon nanotubes forming MCA.

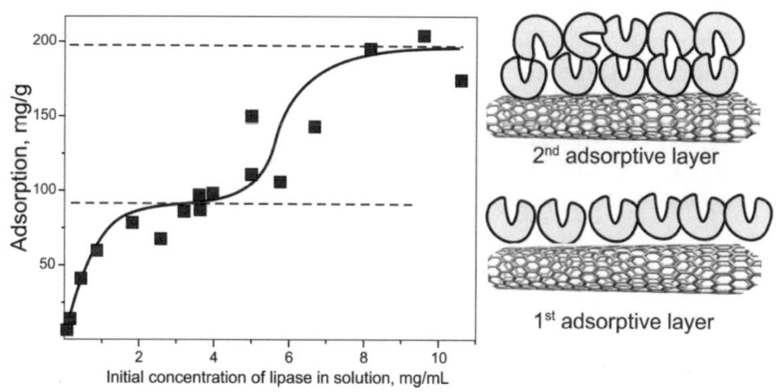

Figure 7.
Adsorption of recombinant lipase on macroporous carbon aerogel depending on initial concentration of soluble lipase. Conditions of adsorption: 20 ± 2°C, 0.02 M Na-phosphate buffer pH 7.0, 24 h.

the 2nd plateau (**Figure 7**). Considering the molecular weight of the lipase, ~30,000 Da, we calculated the surface concentration of the adsorbed lipase, 2×10^{15} molecules per 1 m^2, and then the area occupied by one molecule of the adsorbed lipase, ~500 nm^2. The diameter of one adsorbed molecule of lipase was estimated to be ~20 nm as confirmed by HRTEM images [19]. It is known from literature [22] and PDB-database, that crystallographic size of the lipase molecule is $3.5 \times 4.5 \times 5.0$ nm; in aqueous media, the diameters of hydrated molecules increase twice. The *T. lanuginosus* lipase has the ability to be associated into dimers due to hydrophobic interaction between hydrophobic lids and pockets of active site [23]. So, the first plateau corresponded to the formation of the 1st adsorptive layer that was dense and uniform in distribution of the adsorbed lipase on the surface of MCA. The second plateau can be attributed to the 2nd also dense adsorptive layer formed probably by further dimerization of lipase molecules. As it turned out, under studied conditions, the lipase adsorbed molecules formed two dense adsorptive layers; the amount of lipase in each adsorptive layer was equal to ~100 mg per 1 g of MCA (**Figure** 7).

It was found that 85–90% of the amount of adsorbed lipase was very firmly attached to the surface of carbon aerogel and did not desorb [19]. Desorption was investigated using various selective reagents in order to clarify mechanism of lipase binding on carbon surface. For example, the amount of lipase desorbed using 1 M sodium chloride did not exceed 2%; hence, electrostatic interactions were negligible. The most effective desorption, 10 ± 2%, was observed using distilled water and C2-C4 alcohols' solutions; hence, weak Van der Waals interactions were predominant. The lipase was also efficiently desorbed by emulsifier (gum arabic) perhaps due to breaking hydrophobic interactions. Thus, study on selective desorption showed that the strong adsorption of *T. lanuginosus* lipase on carbon nanotubes forming aerogel occurred exclusively due to hydrophobic-hydrophobic interactions.

Lipase-active biocatalysts prepared by adsorptive immobilization of recombinant *T. lanuginosus* lipase on macroporous carbon aerogel were studied in the periodic processes of bioconversion of triglycerides and fatty acids such as tributyrin hydrolysis, interesterification of vegetable oil with ethyl acetate, and esterification of saturated fatty acids (butyric C4:0, capric C10:0, and stearic C18:0) with *iso*-pentanol.

The activity of the lipase-active biocatalysts and specific activity of adsorbed on MCA lipase were measured in hydrolysis of emulsified tributyrin under conditions when there were no diffusion limitations for mass transfer of the substrate from reaction media toward the adsorbed lipase [19]. Note that specific activity of adsorbed enzyme (in U/mg) is calculated by dividing the experimentally observed activity of the biocatalyst (in U/g) by the amount of adsorbed enzyme (in mg/g). The specific activities of the lipase adsorbed both on MCA granules and nonporous fine powders of carbon nanotubes were compared with each other; the values were found to be 160 and 12 U/mg, respectively. This significant difference was due to the relative rigidity of ball-shaped aerogel granules that prevented agglomeration of carbon nanotubes in "oil-in-water" triglyceride emulsion as presented in **Figure 5b**. Since water-immiscible hydrophobic triglyceride molecules were adsorbed efficiently on hydrophobic carbon nanotubes, with increasing tributyrin concentration, agglomerates of round shape containing this substrate were formed from CNTs (**Figure 5b**). As a result, the required hydrolysis water is displaced from the vicinity of adsorbed lipase and its specific activity fell down as mentioned above.

The maximal hydrolytic specific activity of the adsorbed *T. lanuginosus* lipase was measured to be 700 U/mg (vs. 14,000 U/mg for soluble lipase), i.e., activity significantly decreased upon adsorption of lipase on MCA. Another reason discussed below may be probably incorrect orientation of adsorbed lipase on a highly hydrophobic carbon surface of CNTs.

The results of study on dependence of the specific activity of adsorbed lipase on the adsorption value are presented in **Figure 8a**. Initially, when adsorption of the lipase was small, the observed activities were extremely low. The specific activity increased dramatically and reached maximal values, >700 U/mg, in a region close to the formation of the 1st protein dense layer at adsorption of 100–110 mg/g (**Figure 8a**). Then, the specific activity fell down as adsorption increased (**Figure 8a**). As a result, maximal hydrolytic activity of the biocatalysts, 75,000 U/g, was observed at the adsorption corresponding to the formation of the 1st protein layer. The activity of the biocatalyst with the double adsorptive layer was found to be 1.6-fold lower, 47,000 U/g. In order to explain maximum on the curve in **Figure 8a**, we proposed that the adsorbed lipase can be oriented differently on the

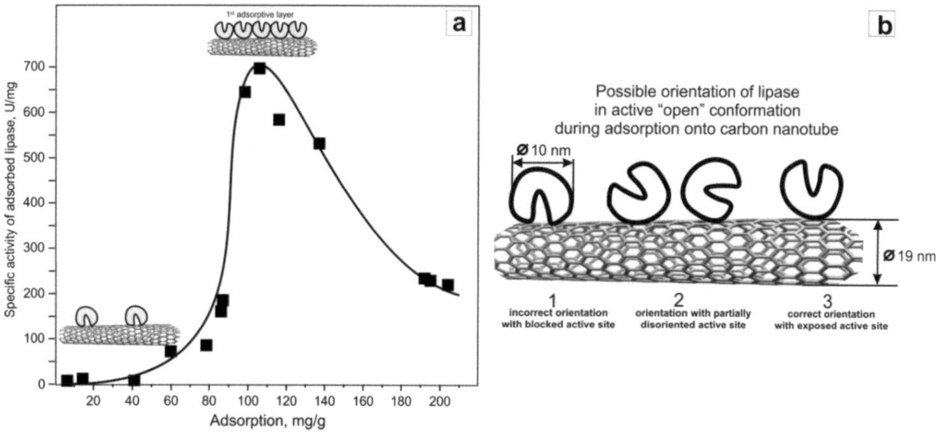

Figure 8.
Specific hydrolytic activity of lipase adsorbed on carbon aerogel depending on adsorption. Conditions of hydrolysis: 20 ± 2°C, 0.02 M Na-phosphate buffer pH 7.0, 0.02 M tributyrin, 1.0 M glycerol, 0.6% gum arabic, content of biocatalyst in reaction medium 3.3 wt.%.

surface of carbon nanotubes, namely correctly and incorrectly (**Figure 8b**). The incorrect orientation (No. 1 in **Figure 8b**) with blocking the active site by the surface was realized more likely in the incompact 1st adsorptive layer; in this case, the observed lipase activity had a minimum value. The correct orientation (No. 3 in **Figure 8b**) may be realized in the dense 1st adsorptive layer, so the formation of active enzyme-substrate complex occurred; and in this case, the observed lipase activity had a maximum value. A further decrease in specific activities after the formation of the 1st adsorptive layer was due to covering the active lipase by the molecules in subsequent 2nd adsorptive layer (scheme in **Figure 7**) and possible partially correct orientation (No. 2 in **Figure 8b**). The obtained data suggest that the main reasons for a loss of enzymatic activity upon adsorption of the lipase on hydrophobic carbon aerogel are as follows: (1) incorrect orientation of enzyme molecule on the carbon surface and blockage of the active site, as well as possible deformation of enzyme molecule; and (2) dehydration of the biocatalysts due to removal of essential water from the vicinity of adsorbed lipase via efficient adsorption of hydrophobic triglycerides.

It has been found that activity and stability of the prepared lipase-active heterogeneous biocatalysts depended strongly on the type of enzymatic reaction performed either in aqueous reaction media (hydrolysis) or in nonaqueous media (interesterification and esterification). The stability of the biocatalysts was quite low when triglycerides participated in reactions as lipase substrates. For example, the biocatalysts lost ~90% of initial activity during six reaction cycles of tributyrin hydrolysis [19]. Another example, the stability of adsorbed on MCA r*Pichia*/lip was determined during interesterification of linseed oil with ethyl acetate in order to produce the valuable product, vitamin F—ethyl esters of ω-3 fatty acids. The conversion of triglycerides was 87 and 66% in the 1st and 3–5th reaction cycles, respectively, and then biocatalysts inactivated due to rapid dehydration.

As mentioned above, esterification of acid with alcohols was accompanied by the formation of water and the corresponding ester, and, as a result, accumulation of essential water molecules inside biocatalysts in the vicinity of the adsorbed lipase occurred. So, the stability of the prepared biocatalysts was much higher in esterification than in hydrolysis and interesterification. And the biocatalysts operated during esterification without a loss of activity for more than several hundred hours in the nonaqueous media of anhydrous organic solvents. As seen in **Figure 9**, the

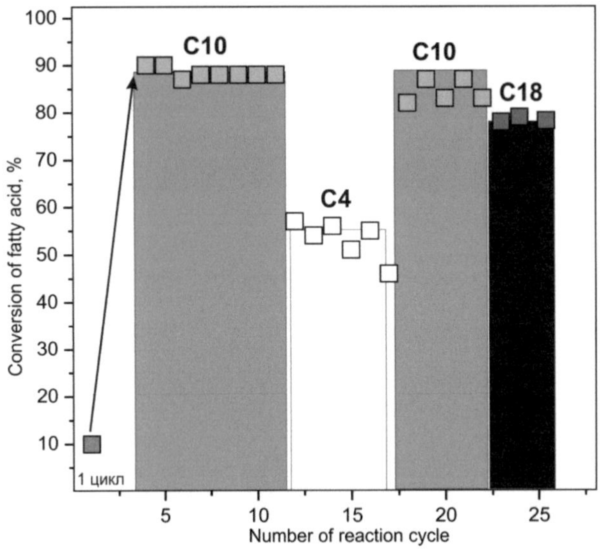

Figure 9.
Operational stability of the biocatalyst in periodic esterification of various fatty acids such as butyric (C4), capric (C10), and stearic (C18) with iso-pentanol as a function of the acid conversion depending on a number of reaction cycle. Conditions of esterification: 40°C, 0.10 M acid, 0.40 M alcohol, hexane:diethyl ether = 1:2, agitation 70 rpm, content of biocatalyst in reaction medium 3.3 wt.%. The amount of adsorbed lipase is 206.0 mg g^{-1}.

esterification activity of dried biocatalysts sharply increased during the 1st reaction cycle; a pre-conditioning stage was observed as described above. The esterified activity depended on the structure of fatty acid, namely of a number of carbon atoms in molecule. Similar to LipoSil biocatalysts described above, the rate of esterification of butyric C4 acid was the lowest compared to capric (C10:0) and stearic (C18:0) acids (**Figure 9**). In the process of the synthesis of *iso*-pentyl caprinate in binary (hexane and diethyl ether) solvent at 40°C, the prepared biocatalysts demonstrated high operational stability and operated without losing activity for a few hundred hours [19].

3.4 Conclusion for the part 3.2

Macroporous carbon aerogel (MCA) obtained by *in situ* catalytic synthesis of the carbon nanotubes was studied as the efficient promising support for adsorptive immobilization of enzymes, in particular recombinant *T. lanuginosus* lipase, due to the rigid 3D-framework of granules and high adsorption ability of MCA. Heterogeneous lipase-active biocatalysts were prepared by adsorption of the enzyme on carbon nanotube (CNT)-forming aerogel, and the processes of bioconversion of substrates into valuable products were organized in periodic regimes under mild ambient conditions.

Recombinant *T. lanuginosus* lipase was adsorbed on macroporous carbon aerogel very firmly exclusively due to hydrophobic interactions between enzyme and carbon nanotubes. Two dense adsorptive layers were formed by lipase molecules; the amount of the lipase in each layer was nearly 100 mg/g.

Activity and stability of the prepared biocatalysts strongly depended on the type of reaction media, namely aqueous or nonaqueous. In the hydrolysis of triglycerides (aqueous reaction medium), the specific activity of the *T. lanuginosus* lipase significantly (in 20 times) decreased upon adsorption on MCA, because orientation of adsorbed lipase on the carbon surface was probably incorrect. The maximal activity of the biocatalyst, 75,000 U g^{-1}, was measured at the adsorption corresponding to the formation of the 1st dense adsorptive layer. The activity and stability of the prepared biocatalysts decreased during the periodic process of tributyrin hydrolysis because a displacement of essential water via efficient adsorption of triglyceride on CNTs occurred, which was followed by progressive dehydration of biocatalysts.

In esterification of fatty acids with alcohol (nonaqueous reaction media), hydrophobic MCA did not prevent the accumulation of the produced essential water in the vicinity of adsorbed lipase. The lipase-active biocatalysts possessed high stability in the synthesis of esters of fatty acids with *iso*-pentanol and operated for a few hundred hours under mild condition of the synthesis of *iso*-pentyl caprinate in binary (hexane and diethyl ether) solvent at 40°C.

As a general conclusion based on all results of adsorption/desorption and activity/stability, we can convincingly note that texture and chemical nature of supports such as mesoporous silica and macroporous carbon aerogel greatly affect the biocatalytic properties of the adsorbed recombinant *T. lanuginosus* lipase. This influence was due to not only hydrophilic-hydrophobic properties of the adsorbents but also specific molecular features of lipase and peculiarities of biocatalysis carried out in aqueous and nonaqueous reaction media. The main reason of inactivation of the prepared biocatalysts during bioconversion of hydrophobic triglycerides, such as hydrolysis and interesterification, was their dehydration. The accumulation of essential water inside the lipase-active biocatalysts during esterification prevented their inactivation. Analyzing our own and literature data, it was concluded that due to the prominent catalytic properties, such as enzymatic activity and operational stability, in combination with simplicity of the immobilization method, the biocatalysts prepared by adsorption of recombinant lipase onto mesoporous silica and macroporous carbon aerogel are promising for practical implementation for reactions of organic synthesis, in particular for production of fatty acids' esters.

Acknowledgements

The authors are grateful to Anatoly Beklemishev and Maria Pykhtina for design strain of r*Pichia*/lip producing recombinant *T. lanuginosus* lipase for this research. This work was conducted within the framework of the budget Project No. AAAA-A17-117041710075-0.

Author details

Galina Kovalenko[1,2]* and Larisa Perminova[1]

1 Institute of Catalysis, Novosibirsk, Russia

2 Novosibirsk State University, Russia

*Address all correspondence to: galina@catalysis.ru

References

[1] Bommarius A, Reibel B. Biocatalysis. Wiley-VCH; 2004. 671 p

[2] Tao J, Kazlauskas R. Biocatalysis for Green Chemistry and Chemical Process Development. Wiley, John Wiley & Sons, Inc.; 2011. 479 p

[3] Bucholdz K, Kasche V, Bornscheuer UT. Biocatalysts and Enzyme Technology. Wiley-VCH; 2005. 448 p

[4] Reilly P. Applied Biochemistry and Bioengineering Enzyme Technologies. In: Wingard LB, Ratchalski-Ratzir T, Golstein L, editors. NY: Academic Press; Vol. 2. 1979. pp. 185-207

[5] Bahar T, Celebi SS. Perfomance of glucoamylase in a magnetically stabilized fluidized bed reactor (MSFBR). Enzyme and Microbial Technology. 2000;**26**:28-33. PII: S0141-0229(99)00129-5

[6] Li X-D, Wu J, Jia D-C, Wan Y-H, Yang N, Qiao M. Preparation of cross-linked glucoamylase aggregates immob ilization by using dextrin and xanthan g um as protecting agents. Catalysts. 2016; **6**:77. DOI: 10.3390/catal6060077

[7] Kovalenko GA, Perminova LV. Immobilization of glucoamylase by adsorption on carbon supports and its application for heterogeneous hydrolysis of dextrin. Carbohydrate Research. 2008;**343**:1202-1211. DOI: 10.1016/j.carres.2008.02.006

[8] Kovalenko GA, Sukhinin SV, Perminova LV. Vortex reactors for heterogeneous biocatalytic processes. In: Krylov IA, Zaikov GE, editors. Industrial Application of Biotechnology. NOVA Science Publisher; 2006. pp. 45-53

[9] Halgas J. Biocatalysts in Organic Synthesis. Elsevier; 1992. 334 p

[10] Hernandez-Martin E, Otero C. Different enzyme requirements for the synthesis of biodeisel: Novozym® 435 and Lipozyme® TL IM. Bioresource Technology. 2008;**99**:277-286. DOI: 10.1016/j. biortech.2006.12.024

[11] Stoytcheva M, Montero G, Toscano L, Gochev V, Valdez B. The immobilized lipase in biodiesel production. In: Stoytcheva M, editor. Biodiesel—Feedstocks and Processing Technologies. 2011. pp. 397-410

[12] Du W, Xu Y, Liu D, Zeng J. Comparative study on lipase-catalyzed transformation of soybean oil for biodiesel production with different acyl acceptors. Journal of Molecular Catalysis B: Enzymatic. 2004;**30**:125-129. DOI: 10.1016/j.molcatb.2004.04.004

[13] Kovalenko G, Beklemishev A, Perminova L, Mamaev L, Moseenkov S, Kuznetsov V. Immobilization of recombinant *E.coli* thermostable lipase by entrapment inside silica xerogel and nanocarbon-in-silica composites. Journal of Molecular Catalysis B: Enzymatic. 2013;**98**:78-86. DOI: 10.1016/j.molcatb.2013.09.022

[14] Perminova L, Kovalenko G, Beklemishev A, Mamaev A, Pykhtina M, Rudina N. Catalytic properties of lipase entrapped as lysate of recombinant strain-producer *Escherichia coli*/lip into nanocarbon-in silica composites in bioconversion of triglyceride and fatty acids. Applied Biochemistry and Microbiology. 2018;**54**:38-44. DOI: 10.1134/S000368381801009X

[15] Perminova L, Kovalenko G, Chukanov N, Patrushev Y. Enzymatic esterification of saturated fatty acids with aliphatic alcohols as an alternative method of a low-temperature synthesis of esters. Russian Chemical Bulletin. 2017;(11):2194-2197

[16] Kovalenko G, Perminova L, Beklemishev A, Mamaev A, Patrushev Y. Biocatalytic heterogeneous processes of the esterification of saturated fatty acids with aliphatic alcohols. Catalysis in Industry. 2018;**10**:68-74. DOI: 10.1134/S2070050418010075

[17] Kovalenko G, Perminova L, Chuenko T, Rudina N. Catalytic properties of lipase adsorbed on nanocarbon-containing mesoporous silica in esterification and interesterifiation reactions. Applied Biochemistry and Microbiology. 2016; **52**:582-588. DOI: 10.1134/S0003683816060089

[18] Kovalenko G, Perminova L, Beklemishev A. Catalytic properties of recombinant *Thermomyces lanuginosus* lipase immobilized by impregnation into mesoporous silica in the enzymatic esterification of saturated fatty acids with aliphatic alcohols. Reaction Kinetics and Mechanisms and Catalysis. 2019;**128**:479-491. DOI: 10.1007/s11144-019-01648-z

[19] Kovalenko G, Perminova L, Krasnikov D, Kuznetsov V. Macroporous carbon aerogel for immobilized enzymes and a support for the lipase-active biocatalysts for conversion of triglyceride and fatty acids. Journal of Porous Materials. 2018; **25**:1017-1026. DOI: 10.1007/s10934-017-0512-0

[20] Turati D, Morais W, Terrasan C, Moreno-Perez S, Pessela B, Fernandez-Lorente G, et al. Immobilization of lipase from *Penicillium* sp. section Gracilenta (CBMAI 1583) on different hydrophobic support: Modulation of functional properties. Molecules. 2017; **22**:339-353. DOI: 10.3390/molecules22020339

[21] Silveria E, Moreno-Perez S, Basso A, Serban S, Mamede R, Tardioli P, et al. Modulation of the regioselectivity of *Thermomyces lanuginosus* lipase via biocatalysts engineering for the ethanolysis of oil in fully anhydrous medium. BMC Biotechnology. 2017;**17**: 88-101. DOI: 10.1186/s12896-017-0407-9

[22] Fernandez-Lafuente R. Lipase from *Thermomyces lanuginosus*: Uses and prospects as an industrial biocatalyst. Journal of Molecular Catalysis B: Enzymatic. 2010;**62**:197-212. DOI: 10.1016/j.molcatb.2009.11.010

[23] Palomo J, Ortiz C, Fuentes M, Fernandez-Lorente G, Guisan J, Fernandez-Lafuente R. Use of immobilized lipases for lipase purification via specific lipase-lipase interactions. Journal of Chromatography A. 2004;**1038**:267-273. DOI: 10.1016/j.chroma.2004.03.058

Chapter 2

Potential of Biocatalysis in Pharmaceuticals

Snehi Soy, Riddhi Prabha and Vinod Kumar Nigam

Abstract

Biocatalysis has been continuously evolving as an essential tool which is playing a significant role in the industrial synthesis of chemicals, active pharmaceuticals, pharmaceutical intermediates, etc. where the high-yielding chemo-, regio-, and enantioselective reactions are needed. Despite its vital importance, industrial biocatalysis is facing certain limitations such as operational stability, economic viability, efficient recovery, and reusability. The limitations mentioned can be overcome by the isolation of specific enzyme producers from extreme environment by protein engineering, bioinformatics, and recombinant DNA technologies. Recently, chemoenzymatic pathway and biological cascade reactions have also been developed and designed to perform the synthesis of pharmaceuticals. In this chapter, we compile the broad applications of biocatalysts in the synthesis of pharmaceuticals.

Keywords: biocatalysis, biocatalyst, enantiomers, pharmaceuticals, substrate specificity, stability

1. Introduction

Biocatalysis is appropriately defined as the enzyme-based applications for the transformation of molecular substrate into several natural as well as synthetic chemicals [1, 2]. The enzymes used in the process are in the form of cell lysate, whole cells, or purified enzyme and are prepared either as recombinant expressed proteins in different host cells or expressed in their native cells itself [3]. The key players of biocatalysis are biocatalysts or enzymes that have been divided into six classes by the IUPAC nomenclature system based on the reactions they catalyze [4], as shown in **Table 1**.

Enzymes as biocatalysts are incredibly proficient and are always preferred to conventional chemical processes. It is due to the fact that enzyme-based biocatalysis has distinct advantages over chemical reactions such as (1) significant specificity towards catalyzed reactions and recognized substrates, (2) simplified synthetic route, (3) high yields with exceptional regio-, chemo-, and stereoselectivities, (4) minimum energy requirements, and (5) generation of less by-products and wastes [5–8]. Another preferred advantage includes whole bioprocess and bulk operations being carried out under mild conditions at elevated rates and with extreme specificity and with minimum environmental and physiological toxicity, thus making them an ideal candidate in the development and improvement of sustainable chemical processes [9–12].

 IntechOpen

Enzyme class	IUPAC code	Catalyzed reactions	Important subclasses
Hydrolases	EC3	Hydrolytic reactions and their reversal	Esterases, glycosidases, lipases, proteases, peptidases, amidases
Oxidoreductases	EC1	Redox reactions	Dehydrogenases, oxidases, oxygenases, peroxidases, reductases
Transferases	EC2	Functional group transformation, addition/ elimination involving C-C, C-N, and C-C bond formation or breakage	C_1-transferases, glycosyltransferases, aminotransferases, phosphotransferases
Lyases	EC4	Elimination reactions	Aldolases, decarboxylases, dehydratase, few pectinases
Isomerases	EC5	Molecular isomerizations	Epimerases, racemases intramolecular transferases
Ligases/synthetases	EC6	Formation of a covalent bond joining two molecules together, coupled to hydrolysis of an ATP or analog	C-C, C-N, C-O, C-S ligases

Table 1.
IUPAC classification of enzymes based on reactions they catalyze.

However, despite holding tremendous potential, biocatalysis has an inevitable pitfall associated with it when extreme conditions of industrial processes are to be considered. An efficient biocatalyst needs to be compatible enough with specific properties such as thermostability, catalytic ability, substrate specificity, and operational stability in turbulent flow regimes, toxic, hazardous solvents, and substrate inhibition [13–21].

Thus, there is a need for the identification and production of stable biocatalysts with broad industrial applicability by exploring and screening novel microbes or identification of new genes with desired properties through the analysis of genes responsible for enzyme production and stability. Further enhancement of the enzyme properties can be done by applying protein engineering tools such as molecular docking, directed evolution, molecular modeling, and process engineering [22–25].

2. Scenario of biocatalysis in pharmaceuticals industries and its pertinent applications

In 1992, Roger Sheldon estimated environmental impact factor (*E* factor) (kg waste/kg product) for several chemical industries, and an *E* factor of 25–>100 was noted in the pharmaceutical industries [26]. Thus, to reduce the harmful impact of pharmaceutical manufacturing processes and making it more sustainable, "green chemistry" has been increasingly adopted. An efficient biocatalytic process encompasses the "12 principles of green chemistry" to an extent which give it an edge over other technologies [27], as shown in **Figure 1**.

In Europe, a project CHEM21 was launched by the collaboration of both government and industries for the implementation of green technology in the chemical and pharmaceutical sectors [28–30]. The project was launched because of the replacement of biocatalysis over chemical in the synthesis of pharmaceuticals involving several redox reactions, chiral amine synthesis, and regio- and stereospecific hydroxylation of abundant compounds [18, 28, 31]. Since then biocatalysis has been

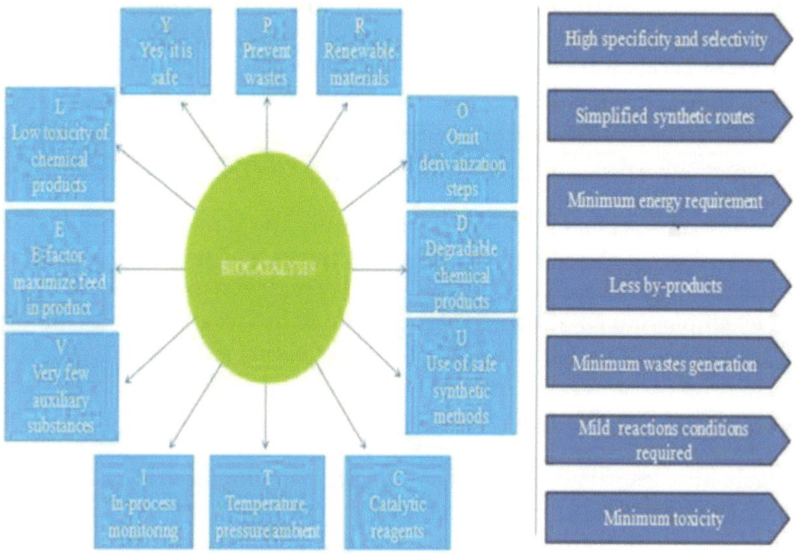

Figure 1.
Schematic representation of biocatalysis benefits embracing the principles of green chemistry.

profitably used for the production of pharmaceutically active chemicals and several blockbuster drugs at the industrial level and some of which are mentioned below:

- Sitagliptin—Sitagliptin, an antidiabetic compound, was successfully produced via biocatalytic approach. It finds application in the treatment of type II diabetes and is sold under the trade name "Januvia" by Merck [32, 33]. This work was accomplished by engineering R-selective transaminase (R-ATA, ATA-117) from *Arthrobacter* species by researchers at Codexis and Merck. The drug produced was having 99.95% enantiopurity even in the presence of 1 M i-$PrNH_2$ with 50% DMSO and at a temperature >40°C [33]. Conventionally, it was prepared using rhodium, a heavy metal as a catalyst. However, on comparing both processes, the biocatalytic method showed a massive reduction in waste as well as the use of heavy metal. Besides this, the overall yield and productivity were increased by 10 and 53% [34]. The R- and S-selective ATA was also used in the production of a variety of drugs such as niraparib and the production of an antagonist of orexin receptor with the formation of inhibitor of JAK kinase pathway [35–38].

- Boceprevir—This is a product of chiral amine synthesis and is marketed by Merck under trade name Victrelis. It is used for the treatment of chronic hepatitis C infections. In the production process, monoamine oxidase (MAO) from fungus *Aspergillus niger* was used for the asymmetrical amine oxidation of bicyclic proline intermediate [39]. The biocatalytic process increased yield by 150%, with an overall reduction in raw materials and side products as waste. At present, engineered monoamine oxidase (MAO) is also used in the production of another hepatitis C drug, telaprevir [34, 40], and various other synthetic drugs such as solifenacin, levocetirizine along with few natural alkaloid products (confine, harmicine, elegance, and leptaflorine).

- Montelukast—Montelukast or Singulair (trade name) is an anti-asthmatic drug marketed by Merck [41]. The engineered keto-reductase (KRED) was used for the production of montelukast, which displayed significant enantioselectivity

(99.9%) and was stable in 70% organic solvent and temperature of 45°C [24]. The biocatalytic method was advantageous in the sense that it omitted the use of hazardous chemical catalyst chlorodiisopinocampheylborane (DIP-CI), which was conventionally used. Several other drugs such as atorvastatin, crizotinib, duloxetine, and phenylephrine were also developed by biocatalytic process using KRED from bacterium *Lactobacillus kefir* [29].

- Atorvastatin—It comes from the statin family and is marketed under the trade name Lipitor by Pfizer. This drug reduces cholesterol levels by inhibiting the synthesis of cholesterol in the liver [42]. Atorvastatin production is also carried out by employing KRED for the production of hydroxy nitrile, an important intermediate. It is a multienzyme process involving glucose dehydrogenase (GDH), KRED, and halohydrindehalogenase (HHDH). Thus, the process is environmentally as well as economically feasible.

- Pregabalin—Pregabalin, a lipophilic GABA (γ-aminobutyric acid) analog, finds use in the treatment of various central nervous system ailments including neuropathic pain, fibromyalgia, epilepsy, and anxiety [43, 44]. Its production was carried out by biocatalytic conversion of *rac*-2-carboxyethyl-3-cyano-5-methylhexanoic acid ethyl ester to 2-carboxyethyl-3-cyano-5-methylhexanoic acid using lipolase. A heat-promoted decarboxylation of 2-carboxyethyl-3-cyano-5-methylhexanoic acid yielded (*S*)-3-cyano-5-methylhexanoic acid ethyl ester, which is a principal known precursor of pregabalin [45]. The mentioned chemoenzymatic synthesis route not only produced increased yields of pregabalin (40–45%) but also eliminated wastes and usage of organic solvent.

- 7-ACA (7-aminocephalosporic acid)—Cephalosporin has been extensively used as semisynthetic antibiotics; it acts on bacterial cell wall (peptidoglycan) synthesis. 7-Aminocephalosporanic acid (7-ACA), the critical intermediate or precursor for the production cephalosporins, is biocatalytically produced by

Biocatalysts	**Microbial sources**	**Pharmaceutical compounds**	**References**
Lipase B	*Candida antarctica*	Reboxetine	[49]
Carbonyl reductase (YlCR2)	*Yarrowia lipolytica*	Statins	[50]
Oxidase	*P. simplicissimum*	Pinoresinol	[51]
Acyltransferase (LovD)	Whole-cell *Escherichia coli* strain overexpressing LovD	Simvastatin	[52, 53]
Engineered cyclohexanone monooxygenase	—	Armodafinil	[54]
(+)-γ-lactamases	*Bradyrhizobium japonicum* USDA 6	Carbovir, abacavir, melogliptin	[5, 55]
Immobilized lipase	*Thermomyces lanuginosus*	Rasagiline mesylate (active ingredient of AZILECT®)	[56]
Expressing tyrosine phenollyase	*Erwinia herbicola* cells	L-DOPA	[57]
E. coli cells expressing cellobiose 2-epimerase	*Caldicellulosiruptor saccharolyticus*	Lactulose	[58]

Table 2.
List of biocatalysts and their microbial source employed for the synthesis of pharmaceutical drugs.

enzymatic deacylation of cephalosporin-C (CPC). A two-step enzymatic process utilizes D-amino acid oxidase (DAAO) and 7-β-(4-carboxybutanamido)-cephalosporanic acid acylase (GLA) for two consecutive reactions. Also, a single-step conversion from CPC to 7-ACA has been reported [46]. It has been successfully applied for the conversion of CPC to 7-ACA at industrial level [47]. Similarly, 6-aminopencillanic acid has been reported for the synthesis of semisynthetic penicillins using penicillin acylase [48].

Some other noteworthy examples and recent progress being made in pharmaceutical synthesis using enzymes from various sources are represented in **Table 2**.

3. Conclusion

Biocatalysis has made a remarkable journey so far and has been successfully applied for the numerous biotransformation processes in several industries. It has benefitted nearly all sectors, particularly chemical and pharmaceuticals. The flourishing development of economically viable and sustainable chemoenzymatic processes highly depends on the broader availability and applicability of enzymes with robust performance irrespective of extreme conditions. Recent surveys have shown that most of the biocatalysts are being used in the synthesis of pharmaceuticals or drugs or intermediates replacing some of the chemical processes, but their stability, selectivity, and specificity are of prime concern.

4. Future prospects

Based on the literature available on the role of biocatalysts in the drug/pharmaceutical synthesis, biocatalysts with improved desired characteristics can be achieved by a multifaceted approach, as shown in **Figure 2**.

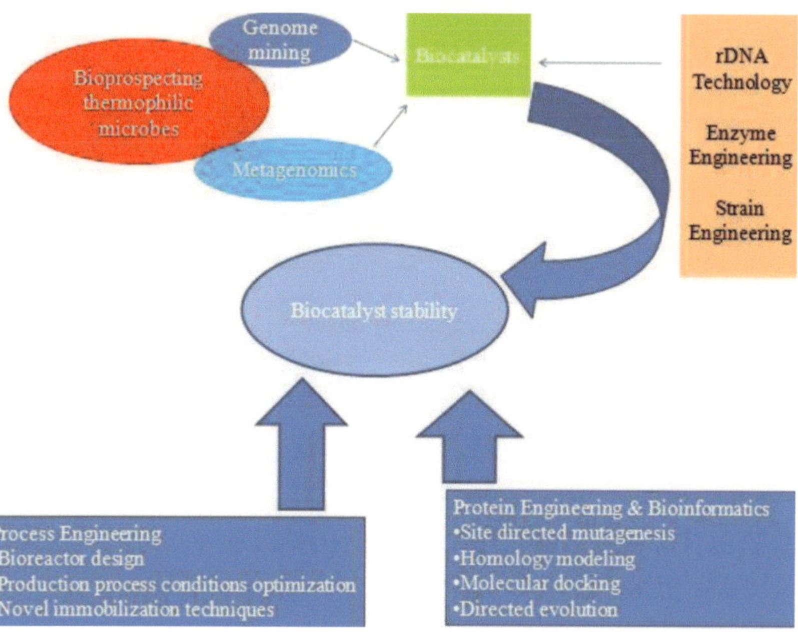

Figure 2.
Schematic representation of improving the operational stability of biocatalyst and enhancing its performance.

Several tools and techniques represented above will enhance the biocatalytic stability, activity, enzyme-substrate affinity, and thermostability and will lead to higher yield. Also, the incorporation of artificial metabolic pathways, cell factory design, and nanotechnology approaches will further aid towards a suitable biocatalytic process. It will also ensure the quality and productivity of the drugs manufactured by optimizing safe process development. Thus, we envision that biocatalysis will be a more radical approach that is going to feat the arena of pharmaceutical manufacturing as well as other sectors such as bioenergy and waste treatment that are far more challenging at present.

Conflict of interest

The authors declare that there are no conflicts of interests whatsoever.

Author details

Snehi Soy, Riddhi Prabha and Vinod Kumar Nigam*
Department of Bio-Engineering, Birla Institute of Technology, Mesra, Ranchi, Jharkhand, India

*Address all correspondence to: vknigam@bitmesra.ac.in

References

[1] Hughes G, Lewis JC. Introduction: Biocatalysis in industry. Chemical Reviews. 2018;**118**:1-3. DOI: 10.1021/acs.chemrev.7b00741

[2] Schmid A, Dordick JS, Hauer B, Kiener A, Wubbolts M, Witholt B. Industrial biocatalysis today and tomorrow. Nature. 2001;**409**:258-268. DOI: 10.1038/35051736

[3] Truppo MD. Biocatalysis in the pharmaceutical industry: The need for speed. ACS Medicinal Chemistry Letters. 2017;**8**:476-480. DOI: 10.1021/acsmedchemlett.7b00114

[4] Sheldon RA, Brady D. Broadening the scope of biocatalysis in sustainable organic synthesis. ChemSusChem. 2019;**12**:2859-2881. DOI: 10.1002/cssc.201900351

[5] Sun H, Zhang H, Ang EL, Zhao H. Biocatalysis for the synthesis of pharmaceuticals and pharmaceutical intermediates. Bioorganic & Medicinal Chemistry. 2018;**26**(7):1275-1284. DOI: 10.1016/j.bmc.2017.06.043

[6] Madhavan A, Sindhu R, Binod P, Sukumaran RK, Pandey A. Strategies for design of improved biocatalysts for industrial applications. Bioresource Technology. 2017;**245**:1304-1313. DOI: 10.1016/j.biortech.2017.05.031

[7] Bommarius AS, Paye MF. Stabilizing biocatalysts. Chemical Society Reviews. 2013;**42**:6534-6565. DOI: 10.1039/C3CS60137D

[8] DiCosimo R, McAuliffe J, Poulose AJ, Bohlmann G. Industrial use of immobilized enzymes. Chemical Society Reviews. 2013;**42**:6437-6474. DOI: 10.1039/C3CS35506C

[9] Woodley JM. Accelerating the implementation of biocatalysis in industry. Applied Microbiology and Biotechnology. 2019;**103**(12):4733-4739

[10] Sheldon RA, Brady D. The limits to biocatalysis: Pushing the envelope. Chemical Communications. 2018;**54**:6088-6104. DOI: 10.1039/C8CC02463D

[11] Sheldon RA, Woodley JM. The role of biocatalysis in sustainable chemistry. Chemical Reviews. 2018;**118**:801-838. DOI: 10.1021/acs.chemrev.7b00203

[12] Ni Y, Holtmann D, Hollmann F. How green is biocatalysis? To calculate is to know. ChemCatChem. 2014;**6**:930-943. DOI: 10.1002/cctc.201300976

[13] Grigoras AG. Catalase immobilization—A review. Biochemical Engineering Journal. 2017;**117**:1-20. DOI: 10.1016/j.bej.2016.10.021

[14] Mehta J, Bhardwaj N, Bhardwaj SK, Kim KH, Deep A. Recent advances in enzyme immobilization techniques: Metal-organic frameworks as novel substrates. Coordination Chemistry Reviews. 2016;**322**:30-40. DOI: 10.1016/j.ccr.2016.05.007

[15] Cao S, Xu P, Ma Y, Yao X, Yao Y, Zong M, et al. Recent advances in immobilized enzymes on nanocarriers. Chinese Journal of Catalysis. 2016;**37**:1814-1823. DOI: 10.1016/S1872-2067(16)62528-7

[16] Cipolatti EP, Valerio A, Henriques RO, Moritz DE, Ninow JL, Freire DMG, et al. Nanomaterials for biocatalyst immobilization—State of the art and future trends. RSC Advances. 2016;**6**:104675-104692. DOI: 10.1039/C6RA22047A

[17] Misson M, Zhang H, Jin B. Nanobiocatalyst advancements, and bioprocessing applications. Interface. 2015;**12**:20140891. DOI: 10.1098/rsif.2014.0891

[18] Choi JM, Han SS, Kim HS. Industrial applications of enzyme biocatalysis:

Current status and future aspects. Biotechnology Advances. 2015;**33**:1443-1454. DOI: 10.1016/j.biotechadv.2015.02.014

[19] Bezerra CS, Lemos CMGDF, Sousa MD, Goncalves LRB. Enzyme immobilization onto renewable polymeric matrixes: Past, present and future trends. Journal of Applied Polymer Science. 2015;**132**:1-15. DOI: 10.1002/app.42125

[20] Ansari SA, Husain Q. Potential applications of enzymes immobilized on/in nano materials: A review. Biotechnology Advances. 2012;**30**:512-523. DOI: 10.1016/j.biotechadv.2011.09.005

[21] Sheldon RA. Enzyme immobilization: The quest for optimum performance. Advanced Synthesis and Catalysis. 2007;**349**:1289-1307. DOI: 10.1002/adsc.200700082

[22] Chapman J, Ismail AE, Dinu CZ. Industrial applications of enzymes: Recent advances, techniques, and outlooks. Catalysts. 2018;**8**(6):238. DOI: 10.3390/catal8060238

[23] Denard CA, Ren H, Zhao H. Improving and repurposing biocatalysts via directed evolution. Current Opinion in Chemical Biology. 2015;**25**:55-64. DOI: 10.1016/j.cbpa.2014.12.036

[24] Bornscheuer UT, Huisman GW, Kazlauskas RJ, Lutz S, Moore JC, Robins K. Engineering the third wave of biocatalysis. Nature. 2012;**485**:185-194. DOI: 10.1038/nature11117

[25] Bornscheuer UT, Pohl M. Improved biocatalysts by directed evolution and rational protein design. Current Opinion in Chemical Biology. 2001;**5**:137-143. DOI: 10.1016/S1367-5931(00)00182-4

[26] Sheldon RA. Organic synthesis; past, present and future. Chemistry and Industry. 1992;**23**:903-906

[27] Sheldon RA. The E factor 25 years on: The rise of green chemistry and sustainability. Green Chemistry. 2017;**19**:18-43. DOI: 10.1039/C6GC02157C

[28] Aldridge S. Industry backs biocatalysis for greener manufacturing. Nature Biotechnology. 2013;**31**:95-96. DOI: 10.1038/nbt0213-95

[29] Huisman GW, Collier SJ. On the development of new biocatalytic processes for practical pharmaceutical synthesis. Current Opinion in Chemical Biology. 2013;**17**:284-292. DOI: 10.1016/j.cbpa.2013.01.017

[30] Tomsho JW, Pal A, Hall DG, Benkovic SJ. Ring structure and aromatic substituent effects on the pKa of the benzoxaborole pharmacophore. ACS Medicinal Chemistry Letters. 2012;**3**:48-52. DOI: 10.1021/ml200215j

[31] Lutz S, Liu LF, Liu YC. Engineering kinases to phosphorylate nucleoside analogs for antiviral and cancer therapy. Chimia. 2009;**63**:737-744. DOI: 10.2533/chimia.2009.737

[32] Desai AA. Sitagliptin manufacture: A compelling tale of green chemistry, process intensification, and industrial asymmetric catalysis. Angewandte Chemie. 2011;**50**:1974-1976. DOI: 10.1002/anie.201007051

[33] Savile CK, Janey JM, Mundorff EC, Moore JC, Tam S, Jarvis WR, et al. Biocatalytic asymmetric synthesis of chiral amines from ketones applied to a sitagliptin manufacture. Science. 2010;**329**:305-309. DOI: 10.1126/science.1188934

[34] Ghislieri D, Turner NJ. Biocatalytic approaches to the synthesis of enantiomerically pure chiral amines. Topics in Catalysis. 2013;**57**:284-300. DOI: 10.1007/s11244-013-0184-1

[35] Chung CK, Bulger PG, Kosjek B, Belyk KM, Rivera N, Scott ME, et al.

Process development of C-N cross coupling and enantioselective biocatalytic reactions for the asymmetric synthesis of niraparib. Organic Process Research & Development. 2013;**18**:215-227. DOI: 10.1021/op400233z

[36] Girardin M, Quellet SG, Gauvreau D, Moore JC, Hughes G, Devine PN, et al. Convergent kilogram scale synthesis of dual orexin receptor antagonist. Organic Process Research & Development. 2013;**17**:61-68. DOI: 10.1021/op3002678

[37] Frodsham L, Golden M, Hard S, Kenworthy MN, Klauber DJ, Leslie K, et al. Use of ω-transaminase enzyme chemistry in the synthesis of a JAK2 kinase inhibitor. Organic Process Research & Development. 2013;**17**:1123-1130. DOI: 10.1021/op400133d

[38] Meadows RE, Mulholland KR, Schurmann M, Golden M, Kierkels H, Meulenbroeks E, et al. Efficient synthesis of (S)-1-(5-fluoropyrimidin-2-yl)ethylamine using an ω-transaminase biocatalyst in a two-phase system. Organic Process Research & Development. 2013;**17**:1117-1122. DOI: 10.1021/op400131h

[39] Li T, Ambrogelly A, Brennan T, Gloor G, Huisman G, et al. Efficient, chemoenzymatic process for manufacture of the bicyclic [3.1.0] proline intermediate based on amine oxidase catalyzed desymmetrization. Journal of the American Chemical Society. 2012;**134**:6467-6472. DOI: 10.1021/ja3010495

[40] Znabet A, Polak MM, Janssen E, de Kanter FJ, Turner NJ, Orru RV, et al. A highly efficient synthesis of telaprevir by strategic use of biocatalysis and multicomponent reactions. Chemical Communications (Camb). 2010;**46**:7918-7920. DOI: 10.1039/C0CC02823A

[41] Liang J, Lalonde J, Borup B, Mitchell V, Mundorff E, Trinh N, et al. Development of a biocatalytic process as an alternative to the (−)-DIP-CI-mediated asymmetric reduction of a key intermediate of montelukast. Organic Process Research and Development. 2010;**14**:193-198. DOI: 10.1021/op900272d

[42] Ma SK, Gruber J, Davis C, Newman L, Gray D, Wang A, et al. A green-by-design biocatalytic process for atorvastatin intermediate. Green Chemistry. 2010;**12**:81-86. DOI: 10.1039/B919115C

[43] de María PD, de Gonzalo G, Alcántara AR. Biocatalysis as useful tool in asymmetric synthesis: An assessment of recently granted patents (2014-2019). Catalysts. 2019;**9**:802. DOI: 10.3390/catal9100802

[44] Rosenthal K, Lutz S. Recent developments and challenges of biocatalytic processes in the pharmaceutical industry. Current Opinion in Green and Sustainable Chemistry. 2018;**11**:58-64. DOI: 10.1016/j.cogsc.2018.03.015

[45] Martinez CA, Hu S, Dumond Y, Tao J, Kelleher P, Tully L. Development of a chemoenzymatic manufacturing process for Pregabalin. Organic Process Research & Development. 2008;**12**:392-398. DOI: 10.1021/op7002248

[46] Nigam VK, Kundu S, Ghosh P. Single-step conversion of cephalosporin-C to 7-aminocephalosporanic acid by free and immobilized cells of *Pseudomonas diminuta*. Applied Biochemistry and Biotechnology. 2005;**126**:13. DOI: 10.1007/s12010-005-0002-8

[47] Tang CD, Shi HL, Jiao ZJ, Shi HF, Yao LG, Xu JH, et al. Exploitation of cold-active cephalosporin C acylase by computer-aided directed evolution and its potential application in low-temperature biosynthesis of

7-aminocephalosporanic acid. Journal of Chemical Technology & Biotechnology. 2018;**93**:2925-2930. DOI: 10.1002/jctb.5647

[48] Hormigo D, López-Conejo MT, Serrano-Aguirre L, García-Martín A, Saborido A, de la Mata I, et al. Kinetically controlled acylation of 6-APA catalyzed by penicillin acylase from Streptomyces lavendulae: Effect of reaction conditions in the enzymatic synthesis of penicillin V. Biocatalysis and Biotransformation. 2019;**13**:1-10. DOI: 10.1080/10242422.2019.1652274

[49] Hayes ST, Assaf G, Checksfield G, Cheung C, Critcher D, Harris L, et al. Commercial synthesis of (s,s)-reboxetine succinate: A journey to find the cheapest commercial chemistry for manufacture. Organic Process Research & Development. 2011;**15**:1305-1314. DOI: 10.1021/op200181f

[50] Xu Q, Tao WY, Huang H, Li S. Highly efficient synthesis of ethyl (S)-4-chloro-3-hydroxybutanoate by a novel carbonyl reductase from *Yarrowia lipolytica* and using mannitol or sorbitol as cosubstrate. Biochemical Engineering Journal. 2016;**106**:61-67. DOI: 10.1016/j.bej.2015.11.010

[51] Ricklefs E, Girhard M, Koschorreck K, Smit MS, Urlacher VB. Two-step one-pot synthesis of pinoresinol from eugenol in an enzymatic cascade. ChemCatChem. 2015;**7**:1857-1864. DOI: 10.1002/cctc.201500182

[52] Xie XK, Watanabe K, Wojcicki WA, Wang CCC, Tang Y. Biosynthesis of lovastatin analogs with a broadly specific acyltransferase. Chemistry & Biology. 2006;**13**:1161-1169. DOI: 10.1016/j.chembiol.2006.09.008

[53] Hoyos P, Pace V, Alcántara AR. Biocatalyzed synthesis of statins: A sustainable strategy for the preparation of valuable drugs. Catalysts. 2019;**9**:260. DOI: 10.3390/catal9030260

[54] Ang EL, Alvizo O, Behrouzian B, et al. Biocatalysts and methods for the synthesis of armodafinil. US20160264945 A1. 2016

[55] Gao S, Zhu S, Huang R, Lu Y, Zheng G. Efficient synthesis of the intermediate of abacavir and carbovir using a novel (+)-γ-lactamase as a catalyst. Bioorganic & Medicinal Chemical Letters. 2015;**25**:3878-3881. DOI: 10.1016/j.bmcl.2015.07.054

[56] Fonseca TdS, Silva MRd, de Oliveira MdCF, Lemos TLGd, Marques RdA, de Mattos MC. Chemoenzymatic synthesis of rasagiline mesylate using lipases. Applied Catalysis A: General. 2015;**492**:76-82. DOI: 10.1016/j.apcata.2014.12.015

[57] Patel RN. Synthesis of chiral pharmaceutical intermediates by biocatalysis. Coordination Chemistry Reviews. 2008;**252**:659-701. DOI: 10.1016/j.ccr.2007.10.031

[58] Wang M, Yang R, Hua X, Shen Q, Zhang W, Zhao W. Lactulose production from lactose by recombinant cellobiose 2-epimerase in permeabilised *Escherichia coli* cells. International Journal of Food Science and Technology. 2015;**50**:1625-1631. DOI: 10.1111/ijfs.12776

Chapter 3

Hydrolase-Catalyzed Promiscuous Reactions and Applications in Organic Synthesis

Yun Wang and Na Wang

Abstract

The potential of biocatalysis becomes increasingly recognized as an efficient and green tool for modern organic synthesis. Biocatalytic promiscuity, a new frontier extended the use of enzymes in organic synthesis, has attracted much attention and expanded rapidly in the past decade. It focuses on the enzyme catalytic activities with unnatural substrates and alternative chemical transformations. Exploiting enzyme catalytic unconventional reactions might lead to improvements in existing catalysts and provide novel synthesis pathways that are currently not available. Among these enzymes, hydrolase (such as lipase, protease, acylase) undoubtedly has received special attention since they display remarkable activities for some unexpected reactions such as aldol reaction and other novel carbon-carbon and carbon-heteroatom bond-forming reactions. This chapter introduces the recent progress in hydrolase catalytic unconventional reactions and application in organic synthesis. Some important examples of hydrolase catalytic unconventional reactions in addition reactions are reviewed, highlighting the catalytic promiscuity of hydrolases focuses on aldol reaction, Michael addition, and multicomponent reactions.

Keywords: enzyme, biocatalysis, promiscuity, hydrolases, lipase, aldol reactions, Michael addition, multicomponent reactions

1. Introduction

Biocatalysis is the application of enzymes for chemical transformations of organic compounds. Enzymes as biocatalysts have many advantages [1–3]: (1) enzymes are very efficient catalysts. Typically the rates of enzyme-mediated processes are accelerated, compared to those of the corresponding nonenzymatic reaction, by a factor of 10^8–10^{10}. The acceleration may even exceed a value of 10^{12}, which is far above the values that chemical catalysts are capable of achieving; (2) enzymes are environmentally acceptable. Unlike heavy metals, for instance, biocatalysts are completely degraded in the environment; (3) enzymes act under mild conditions. Enzymes act in a temperature range of 20–40°C, under neutral aqueous, and in the absence of substrate functional group protection. This minimizes problems of undesired side reactions such as decomposition, isomerization, racemization, and rearrangement, which often plague traditional methodology; (4) enzymes display high chemoselectivity, regioselectivity, and enantioselectivity.

As a result, reactions that generally tend to be "cleaner" and laborious can largely be omitted; and (5) enzymes can catalyze a broad spectrum of reactions. There is an enzyme-catalyzed process equivalent to almost every type of organic reaction, such as oxidation, hydrolysis, addition, halogenation, alkylation, and isomerization. In addition, many enzymes accept unnatural substrates, and genetic engineering can further alter their stability, broaden their substrate specificity, and increase their specific activity. Thus, the application of enzymes in synthesis thus represents a remarkable opportunity for the development of industrial chemical and pharmaceutical processes [4–7].

Although it is well known that a given enzyme is able to catalyze a specific reaction efficiently, some unexpected experimental results have indicated that many enzymes have catalytic promiscuity [8–12]. Enzyme promiscuity is classified into three categories: (a) condition promiscuity, which is an enzyme's ability to work under unexpected condition; (b) substrate promiscuity, which is an enzyme's ability to work with unexpected substrates; and (c) catalytic promiscuity, which is an enzyme's ability to catalyze unexpected reactions. Among them, catalytic promiscuity has gained much attention as it opens a wide scope for the industrial application of enzymes.

During the past decade, biocatalytic promiscuity, as a new frontier extending the use of enzymes in organic synthesis, has received considerable attention and expanded rapidly. A classic example of promiscuous enzymatic behavior is pyruvate decarboxylase, which not only decarboxylates pyruvate but also links acetaldehyde and benzaldehyde to form R-phenylacetylcarbinol. The use of pyruvate decarboxylase to form carbon–carbon bonds, which does not occur in the natural reaction, was first studied in 1921 and was applied in industry today [13]. As one of the most rapidly growing areas in enzymology, multifunctional biocatalytic reactions not only highlights the existing catalysts but may provide novel and practical synthetic pathways which are not currently available. Miao et al. reviewed enzyme promiscuity for carbon-carbon bond-forming reactions like aldol couplings, Michael(-type) additions, Mannich reactions, Henry reactions, and Knoevenagel condensations [14]. Gotor-Fernández et al. also highlighted the hydrolase-catalyzed reactions for nonconventional transformations in the same year [15].

Hydrolases (such as lipase, protease, acylase) have received extensive attention as biocatalysts for a long time due to their many attractive properties like stability in

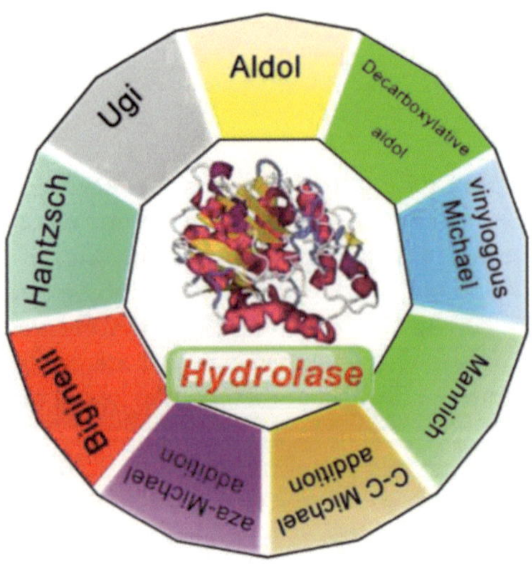

Figure 1.
Hydrolase-catalyzed promiscuous reactions.

organic solvents, not requiring cofactors, broad substrate tolerance, commercial availability, and high chemo-, regio-, and stereoselectivity. Hydrolases have demonstrated a great versatility in hydrolysis, transesterification, aminolysis reactions, etc. Some hydrolase-catalyzed promiscuous reactions have been done in the last decades (**Figure 1**). These research and other relevant reports encouraged us to believe that the catalytic activities for unconventional reactions rather than the well-known hydrolytic function may also have a natural role in hydrolase evolution.

The aim of the present chapter is to give a brief overview of the hydrolase-catalyzed C—C and C—N reactions and present some of the most recent applications in different fields for recent decade. The main work in our group will be disclosed, highlighting the catalytic properties of hydrolases to catalyze not only single processes but also multicomponent and tandem reactions. Consequently, the promiscuous hydrolase-catalyzed reactions are outlined with focus on Michael addition, aldol reaction, Mannich reaction, Biginelli reaction, etc.

2. Michael addition

Michael addition is a 1,4-addition of a nucleophile to α,β-unsaturated compounds, and it is one of the most fundamental and important reactions for the formation of carbon-carbon bonds and carbon-heteroatom bonds in organic synthesis. Michael addition reactions are traditionally catalyzed under strong basic or acidic conditions, which can cause unwanted side reactions such as further condensation or polymerization of α,β-unsaturated compounds. Thus, biocatalysis can afford a green and facile method for organic synthesis. Among different biocatalysts, hydrolases such as protease and lipase have been widely used as a green and efficient catalyst for Michael addition.

2.1 Carbon-heteroatom bond formation Michael addition

Michael addition is the early promiscuous reaction catalyzed by hydrolase. In 1986, Kitazume et al. reported the hydrolytic enzyme-catalyzed stereospecific Michael addition reactions in buffer solution (pH = 8.0) at 40–41°C (**Figure 2**) [16]. This discovery overthrows the long erroneous concept of enzymology that "biocatalysis must be carried out in aqueous solution," making many organic reactions that cannot be carried out in water be completed in organic solvents and greatly expanding the application scope of enzymes as catalysts. Moreover, enzymes are frequently more stable in organic solvents than in water. Thus, some research groups began to focus on enzyme-catalyzed Michael addition reactions in organic solvents.

Lin and Gotor et al. firstly reported the hydrolase-catalyzed Michael addition of imidazole with acrylates catalyzed by alkaline protease from *Bacillus subtilis* in organic solvent in 2004 [17, 18]. Subsequently, other hydrolase-catalyzed Michael addition reactions were reported. In 2010, Bhanage et al. developed an efficient protocol for the regioselective aza-Michael addition of amines with acrylates using

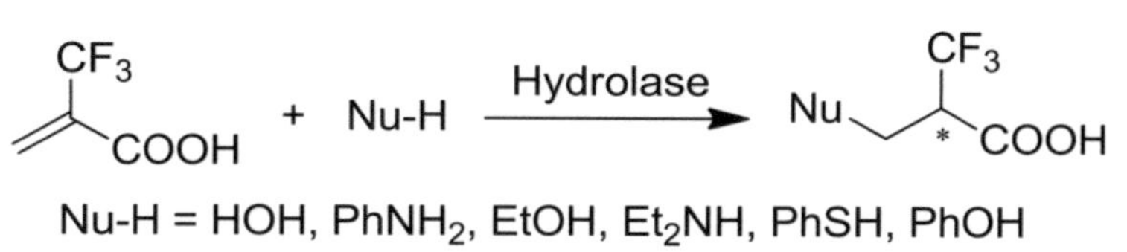

Figure 2.
The first hydrolase-catalyzed Michael addition in buffer.

Candida antarctica lipase B (CALB) as a biocatalyst at 60°C (**Figure 3**) [19]. The universality of the reaction, including the reactions of various primary and secondary amines with different acrylates, was also studied. Higher yields were obtained.

Gotor et al. have explored new synthetic possibilities of commercially available protease from *Bacillus licheniformis* immobilized as cross-linked enzyme aggregates (Alcalase-CLEA) in 2011, since the CLEA immobilization improves the stability of the protein toward denaturalization by heating, organic solvents, and autoproteolysis [20]. Alcalase-CLEA has achieved the best results in the aza-Michael addition of secondary amines to acrylonitrile (**Figure 4**). In all cases the formations of the corresponding Michael adduct were faster than in the absence of biocatalyst, but also in comparison with the inhibited enzyme, the reaction rates being highly dependent of the amine structure.

In 2012, Baldessari et al. firstly reported the synthesis of N-substituted β-amino esters by application of *Rhizomucor miehei* lipase in aza-Michael addition of mono- and bifunctional amines to α,β-unsaturated esters [21]. The authors selected ethyl acrylate and propyl acrylate as the Michael acceptors and different alkylamines, alkanolamines, and diamines as the Michael donors (**Figure 5**). The reactions were carried out in low-polarity solvents (hexane, toluene, and diisopropyl ether (DIPE)) at 30°C for 16 h with yields from 12 to 100%. Subsequently, the authors investigated the effect of the reaction conditions on the Michael addition systematically, such as commercially available enzyme sources, organic solvents, and the structure of the Michael acceptor and donor. The results showed that the alkanolamines in n-hexane were not selective and double Michael adducts could be obtained. Substrate concentration also plays an important role in enhancing the catalytic effect of enzymes on spontaneous reactions. High substrate concentration limits the efficiency of biocatalysts.

In 2013, Demeunynck et al. have optimized the lipase-biocatalyzed addition of benzylamine to ethyl propiolate. Immobilized *Candida antarctica* lipase B was

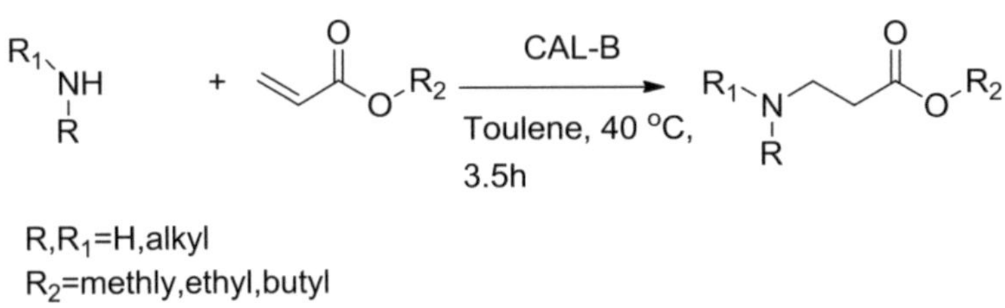

Figure 3.
CALB-catalyzed aza-Michael addition of amine to acrylate.

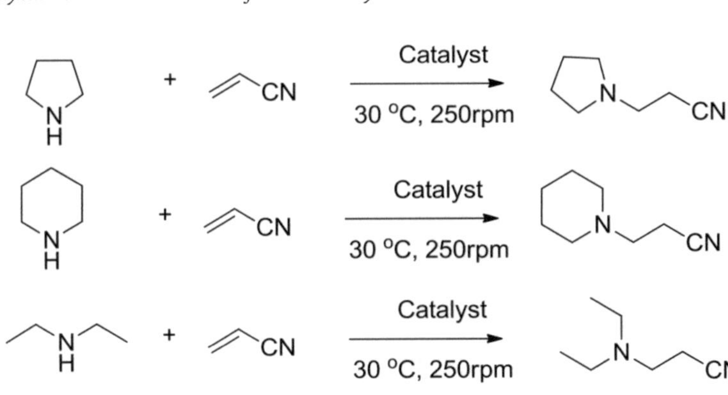

Figure 4.
Hydrolase-catalyzed Michael-type additions between secondary amines and acrylonitrile.

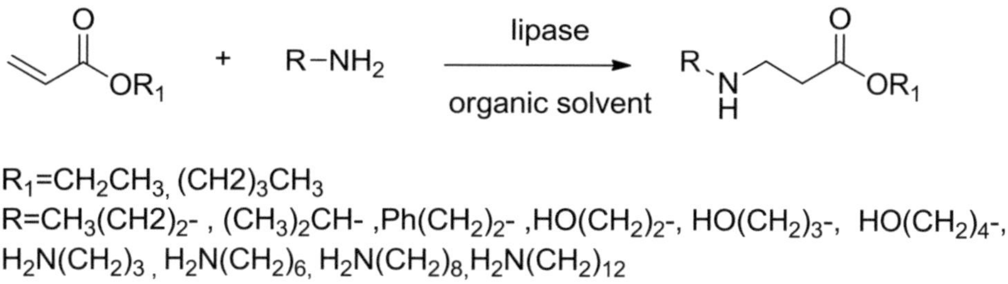

Figure 5.
Lipase-catalyzed aza-Michael addition of amines to α,β-unsaturated esters.

beneficial to the chemoselective 1,2-addition, using TBME, dioxane, or toluene under overnight (15 h) gentle magnetic stirring at 50°C (**Figure 6**) [22]. Under these conditions, the yield of acrylamide was good at the Gram scale. S-trans-z and e-diphenylamine were formed as by-products. The reactions worked well with other primary amines but not with secondary amines that only gave the corresponding aminoacrylates. The chemoselectivity of CALB with N- and S-nucleophiles was also checked. The transesterification also worked in good yields in TBME, toluene, or dioxane. The best yields (near quantitative) were observed when the reactions were carried out in open vessels under gentle magnetic stirring at 50°C for 6 h.

In the same year, Franssen and co-workers demonstrated lipases from *Pseudomonas stutzeri* (PSL) and *Chromobacterium viscosum* (CVL) are excellent catalysts for the aza-Michael addition of amines to substituted or unsubstituted acrylates with high product selectivity and good yields (**Figure 7**) [23]. Comparative studies of other lipases, including Novoxin 435, have proven ineffective. The selective Michael addition of diamines to these substituted acrylates was also realized. In this paper, the catalytic effects of various lipases on aza-Michael addition reaction, especially on the lipase catalysis of *Pseudomonas aeruginosa* OM2 and PSL, are introduced. The 1,4-adducts of acrylate and benzylamine have high yield and selectivity.

Chemoselective synthesis of N-protected β-amino esters involving lipase-catalyzed aza-Michael additions is mainly hampered by the two electrophilic sites present on these compounds. In order to control the chemoselectivity, a solvent engineering strategy based on the thermodynamic behavior of products in media of different polarity was designed by Castillo et al. (**Figure 8**) [24]. This strategy can

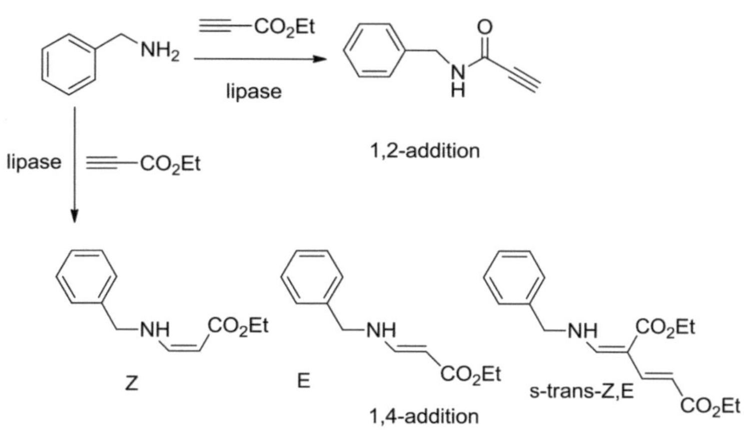

Figure 6.
Reactivity of benzylamine with ethyl propiolate.

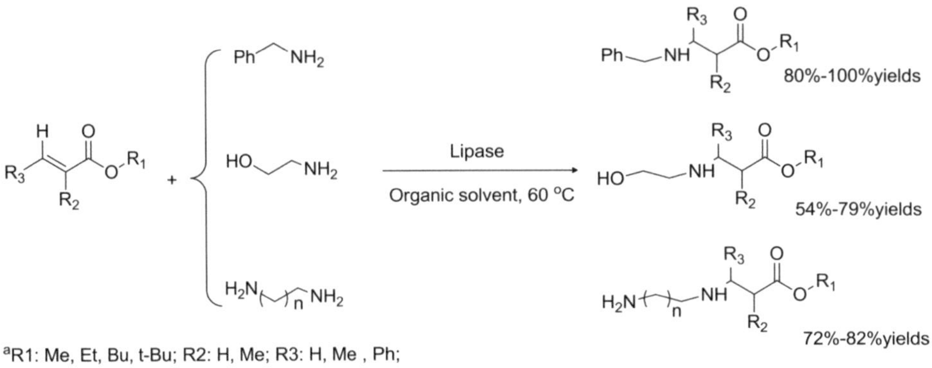

aR1: Me, Et, Bu, t-Bu; R2: H, Me; R3: H, Me , Ph;

Figure 7.
Lipase-catalyzed aza-Michael addition of amines with different substituted acrylates.

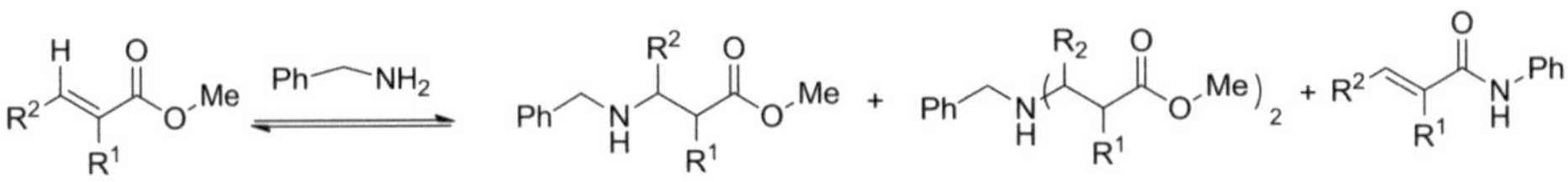

1.R^1=H, R^2=H; 2.R^1=H, R^2=Me; 3.R^1=Me, R^2=H; 4.R^1=Ph, R^2=H; 5.R^1=H, R^2=Ph;

Figure 8.
Chemoselectivity of the addition of benzylamine to α,β-unsaturated esters.

obtain highly selective aza-Michael adducts from benzylamine and different acrylates. Ammonia hydrolysis is avoided in almost all reactions with n-hexane (a nonpolar solvent) as solvent, while the corresponding Michael adduct is synthesized in 53–78% yield. On the contrary, if the reaction is carried out in polar solvents (e.g., 2-methyl-2-butanol (2M2B)), the product of ammonia hydrolysis will be advantageous. Thermodynamic analysis of these processes using the actual solvation conductor-like screening model (COSMO-RS) helps to understand some key factors affecting chemical selectivity and confirms that reliable estimates of the thermodynamic interactions between solutes and solvents allow adequate selection of reaction media that may lead to chemical selectivity.

Reaction media has an important effect on the yield and chemo- and enantioselectivity of biocatalytic reaction. Solvent engineering is an effective tool to direct chemo- and enantioselectivity of the aza-Michael addition and the subsequent kinetic resolution of the Michael adduct [25]. In the reaction of benzylamine and methyl crotonate catalyzed by CALB, three possible adducts can be isolated: aminolysis product, aza-Michael addition product, and double addition product (**Figure 9**). The authors selected n-hexane and 2-methyl-2-butanol (two solvents of opposite polarity) as solvents in the experiments. The Michael adduct is favored in more hydrophobic media, while the amide product in more polar solvents, and the best values of ee for aza-Michael adduct were obtained in almost 100% 2M2B. The experiment results clarify the origin of the enantiomeric excess of the aza-Michael

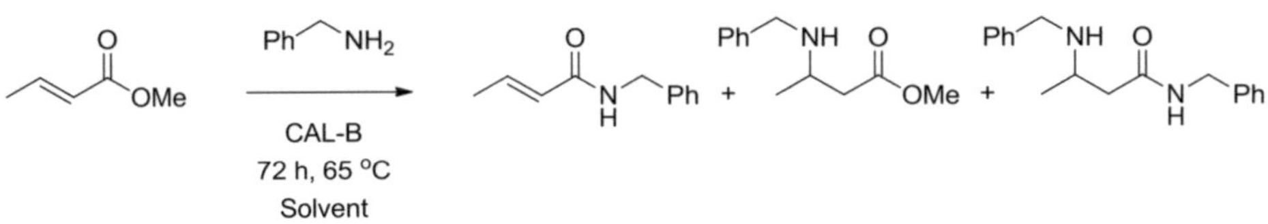

Figure 9.
Lipase-catalyzed addition and aminolysis reaction of benzylamine to methyl crotonate.

addition product, obtained in 2M2B, by a resolution process with CALB on the enantiomers of aza-Michael addition product.

Our group demonstrated that 3-substituted 2H-chromene derivatives were synthesized via biocatalytic domino oxa-Michael/aldol condensations (**Figure 10**) [26]. α-Amylase from *Bacillus subtilis* shows excellent catalytic activity and exerts good adaptability to different substrates in the reaction. The reaction conditions including organic solvents, water content, temperature, molar ratio of substrates, and enzyme loading were optimized.

It is generally believed that the hydrolysis site of hydrolase is also the active site of its miscible catalysis. On this basis, the possible mechanism of domino reaction catalyzed by hydrolase was proposed. First, salicylaldehyde was activated by amino residues and oxygen anions of amylase. Then methyl vinyl ketone is attacked by activated salicylaldehyde, and a new C—O bond is formed by oxa-Michael addition reaction. Next, an intramolecular aldol reaction begins to form carbon–carbon bonds. Finally, the adduct was dehydrated, and the required product was released (**Figure 11**).

Very recently, our group conducted an aza-Michael addition of aniline compounds and acrylate derivatives catalyzed by CALB and several mutants in order to investigate reaction mechanistic (**Figure 12**) [27]. The influence factors of the reaction were discussed systematically, including solvent, enzyme loading, temperature, and time of reaction. On this basis, dozens of substrates with different structures were conducted to occur aza-Michael addition on the optimized conditions. The results demonstrated that the structures of substrates had a great influence on the activity.

Four different reaction intermediates (Intermediate 1, 2, 3, and 4) were matched with the catalytic activity site of CALB to perform molecular docking simulation (**Figure 13A**). We can see that the binding mode of all the four intermediates with the active site is basically the same. The binding modes of four intermediates with CALB catalytic active sites were analyzed, in order to further study the blinding modes of aza-Michael addition intermediates and CALB and the driving forces of their mutual recognition. As shown in **Figure 13b**, it can be seen from the figure

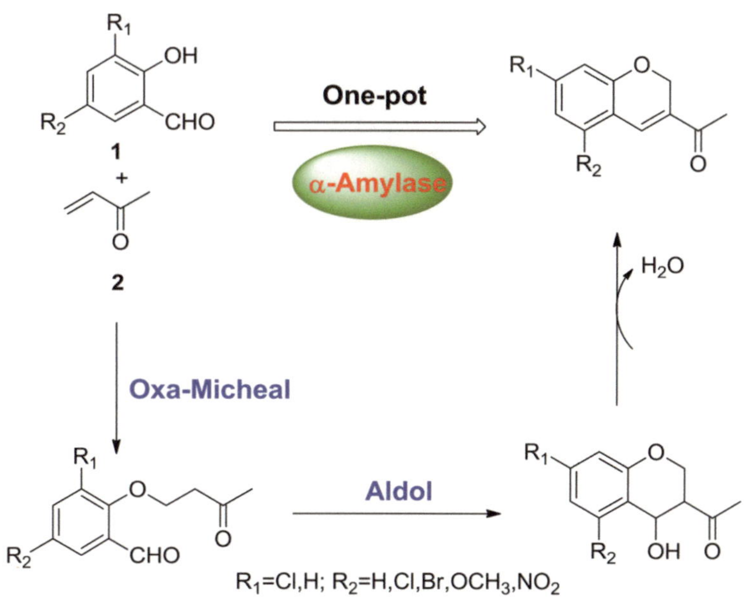

Figure 10.
α-Amylase-catalyzed synthesis of 3-substituted 2H-chromene derivatives.

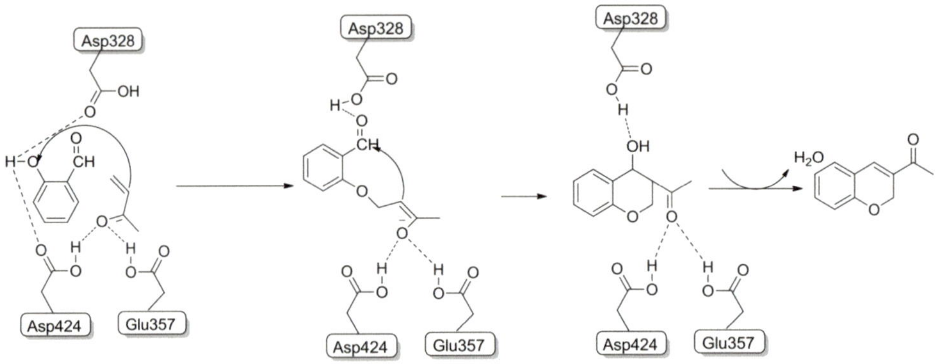

Figure 11.
The proposed mechanism for the α-amylase-catalyzed synthesis of 3-substituted 2H-chromene derivative.

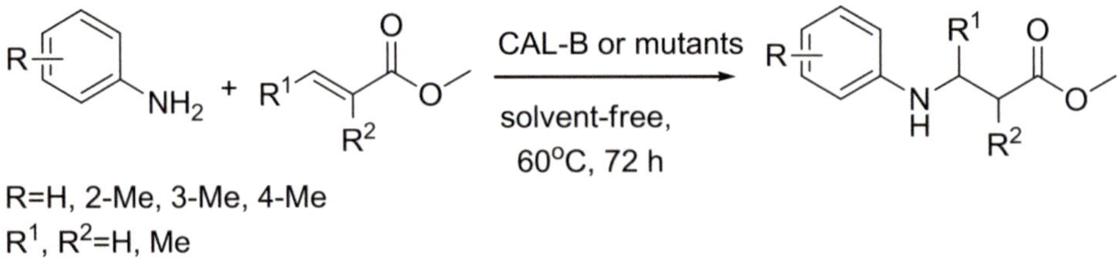

Figure 12.
CALB and mutants catalyzed aza-Michael addition.

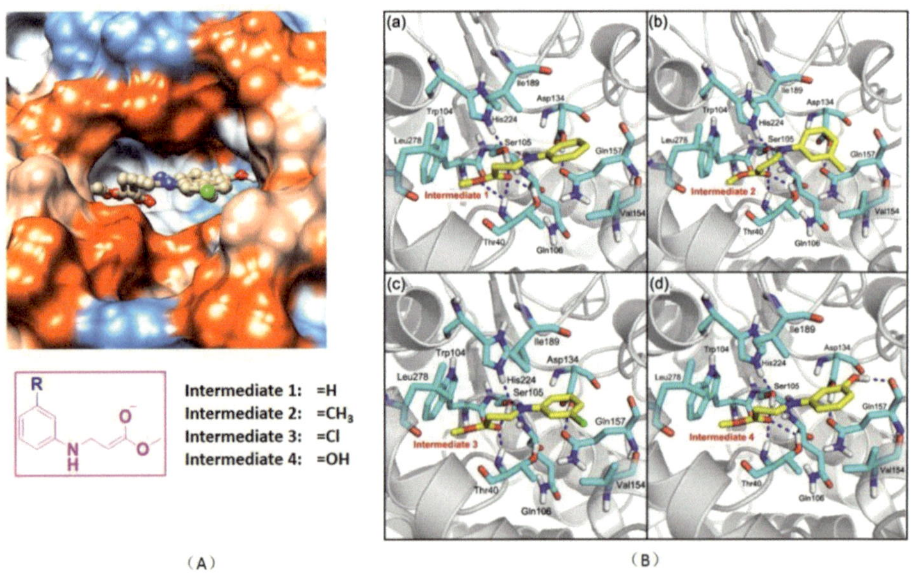

Figure 13.
Molecular docking simulation of CALB with four different reaction intermediates. (A)Hydrophobic matching of the four reaction intermediates with CAL B. The cavity represents the CAL B catalytic pocket which is able to bind and orient the substrates. The blue surface represents hydrophilic while the orange surface represents hydrophobic. The substrates are shown in the pocket in ball-and-stick representation with the atom of substrate coloured according to their atom types (carbon, grey; nitrogen, blue; oxygen, red; chlorine, green). (B)Three-dimensional model of the binding between four aza-Michael addition intermediates (1, 2, 3 and 4) and the CAL B active site. The protein is shown in grey with interacting residues shown as a sky blue stick model. The intermediate is shown as a yellow stick model, and the blue dotted lines indicate the hydrogen bonds between the intermediate and the active site of CAL B.

that the binding modes of the four intermediates with the CALB catalytic chamber are basically the same.

In order to determine the catalytic activity of CALB, three mutants and wild-type CALB were expressed in *E. coli* and were purified to catalyze the aza-Michael addition reaction. The results showed that aza-Michael activity could be dramatically decreased by the mutation of active sites: neither mutant S105 A nor mutant H224 could catalyze the reaction.

However, the mutant I189 A still had a weak catalytic effect on this reaction. Based on these experimental results, the molecular docking was carried out, and the mechanism of aza-Michael addition catalyzed by CALB was studied, and a reasonable reaction mechanism was proposed (**Figure 14**). This helped to explain the effect of substrate structure on the reaction. The substituents of substrates affect the interaction with CALB active sites. Some substituents enhance the binding of substrates and facilitate the reaction. In the whole process, the Ser105 and His224 residues played an important role in proton transfer. Without these two residues, the proton transfer would be blocked, and the aza-Michael addition could not be possible. Besides, the Ile189 residue forms hydrophobic interaction with the benzene ring of the substrate, which makes the substrate more stable in the active cavity.

The biocatalytic thia-Michael reaction is an attractive strategy to develop C—S bond-forming reactions. In 2012, Kiełbasinski and co-workers have reported the use of a number of lipases including PPL, MJL, CALB, and PSL in the addition of benzenethiol to racemic phenyl vinyl sulfoxide or 2-phosphono-2,3-didehydrothiolane S-oxide in organic solvents at room temperature (**Figure 15**) [28]. The addition of piperidine to phenyl vinyl sulfoxide in chloroform is carried out in both enzymatic and non-catalytic processes, while in the former, the reaction rate is 2.5 times faster. Conversely, the conjugate addition of phenylmercaptan with phenyl vinyl sulfoxide is only carried out in the presence of enzymes and ethanol as solvent. In any case, the product is not enantiomerically enriched. However, in the

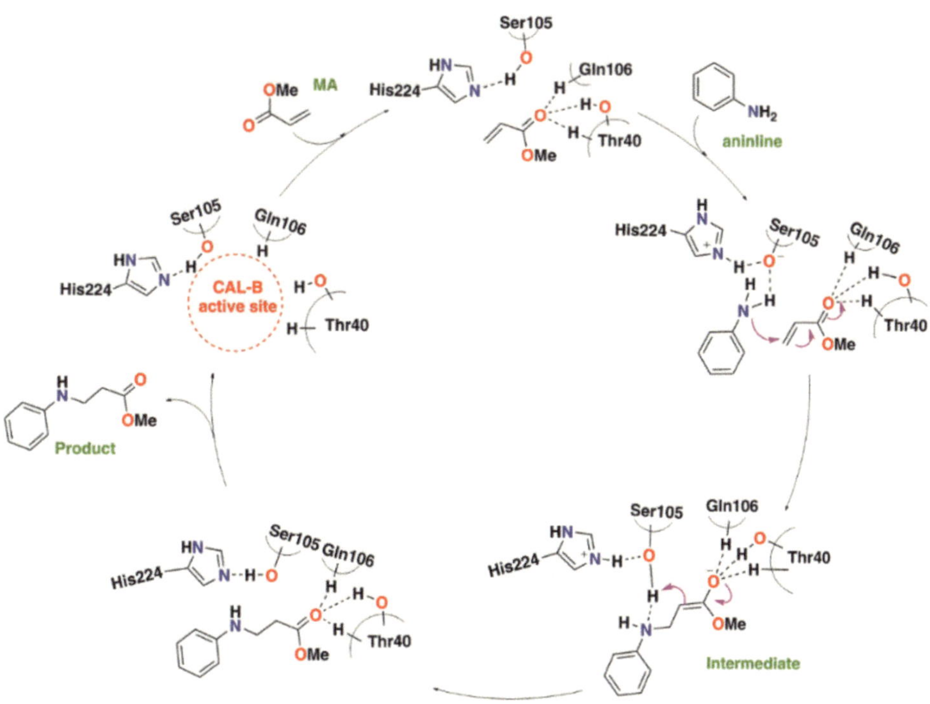

Figure 14.
Proposed reaction mechanism of aza-Michael addition catalyzed by CALB.

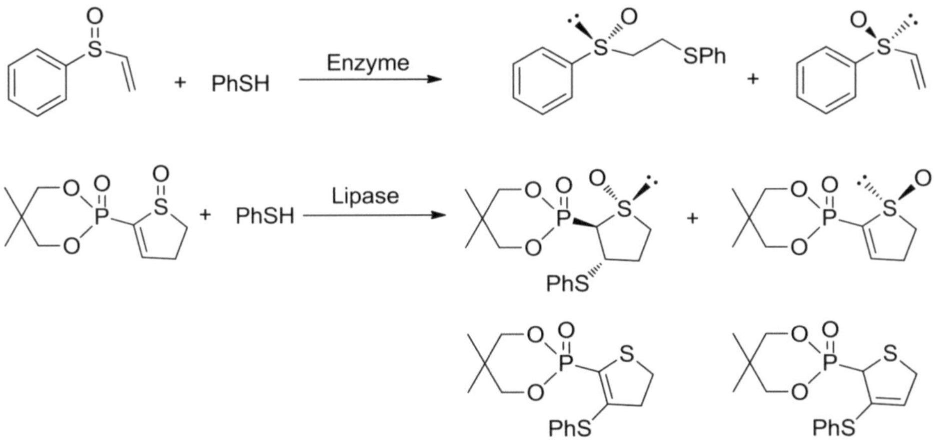

Figure 15.
The Michael addition of benzenethiol to racemic phenyl vinyl sulfoxide or 2-phosphono-2,3-didehydrothiolane S-oxide.

presence of various lipases, the addition of phenylmercaptan to a better Michael receptor, cyclic sulfonyl alkylphosphonate, in some cases resulted in up to 25% optical purity of the product and the recovered substrate.

Then, some mechanistic considerations are presented in the studies. The authors proposed sulfoxide oxygen atoms are bound to the "oxygen anion pore" of the enzyme activity site by hydrogen bond. Conversely, histidine catalyzed by binary enhances the nucleophilicity of sulfur centers in phenylmercaptan molecules. Although the interaction of the latter is the same as Michael's addition of mercaptan to enols, the H-binding of sulfoxide oxygen atom must be different from that of carbonyl oxygen atom, which results in the lower catalytic efficiency of the enzyme for the reaction. It is well known that oxygen anion holes bind to the transition state better than the ground state. When lipase catalyzes ester hydrolysis, the intermediate oxygen anion is tetrahedral. Although the sulfoxide group is tetrahedral, which indicates that the bonding of sulfoxide group should be uniform, compared with the oxygen anion, the sulfoxide group has no negative charge on the intermediate oxygen atom, which significantly reduces the strength of hydrogen bond. In addition, for the Michael addition of nucleophilic reagents, the intermediate oxygen anion is planar, which reduces the space requirement and makes it more suitable for the oxygen anion pore than the tetrahedral sulfoxide intermediate (**Figure 16**).

In 2014, Domingues and co-workers firstly reported the reaction between cinnamaldehyde and thiophenol. Several hydrolases such as PPL, lipozyme, chymosin, and papain have demonstrated different levels of activities, and PPL has found application on the multigram scale (**Figure 17**) [29]. These reactions were carried out at room temperature, and good or excellent sulfur Michael adducts were obtained. The scheme describes the use of EtOH as a solvent and fewer enzymes. The chymosin and papain were used as biocatalysts for organic reactions for the first time.

2.2 Carbon-carbon bond formation Michael addition

C—C bond-forming reactions are one of the mainstays of organic chemistry. In this field the hydrolase-catalyzed Michael reaction also has numerous applications in synthetic chemistry.

In 2011, the asymmetric C—C Michael addition catalyzed by lipozyme TLIM (immobilized lipase from *Thermomyces lanuginosus*) in organic medium in the presence of water was reported for the first time by Guan et al. The biocatalytic reaction

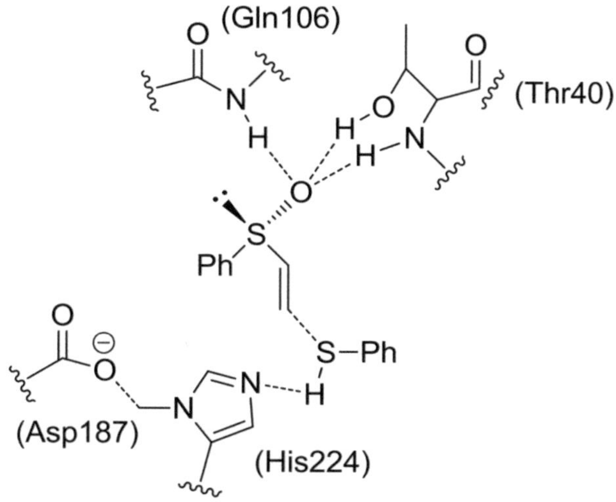

Figure 16.
Assumed mechanism of the conjugate addition of benzenethiol to racemic phenyl vinyl sulfoxide.

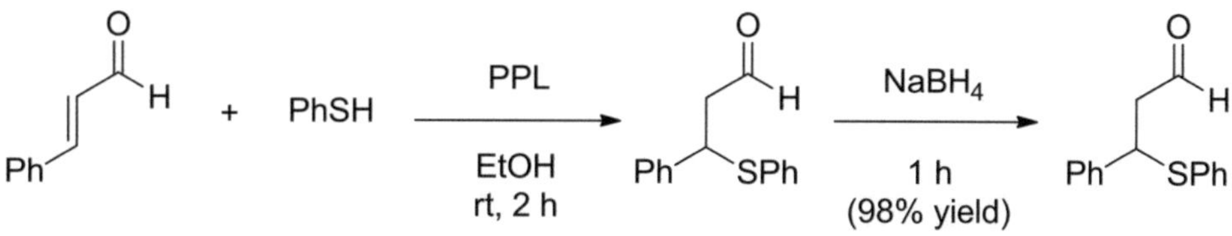

Figure 17.
PPL-catalyzed C—S bond-forming reaction between cinnamaldehyde and thiophenol.

is suitable for adding large amounts of 1,3-dicarbonyl compounds and cyclohexanone to aromatic and heteroaromatic nitroolefins and cyclohexenone in DMSO in the presence of water under mild reaction conditions (**Figure 18**) [30]. The enantioselectivities up to 83% ee and yields up to 90% were achieved.

Then, the same research group explored porcine pancreatic lipase (PPL) which was used as a biocatalyst to catalyze the Michael addition of 4-hydroxycoumarin to α,β-unsaturated enones in organic medium in the presence of water to synthesize warfarin and derivatives in 2012 (**Figure 19**) [31]. The products were obtained in moderate to high yields (up to 95%) with none or low enantioselectivities (up to 28% ee). The influence of reaction conditions including solvents, temperature, and molar ratio of substrates was systematically investigated. It was the first time warfarin and derivatives were prepared using a biocatalyst.

Sometimes Michael adducts are not the final targeted compounds. We studied lipase-catalyzed Michael addition between nitrostyrene and acetylacetone in DMSO in the presence of water under mild reaction conditions in 2011 [32]. Two possible adducts can be isolated: the routine Michael addition product and cyclic product

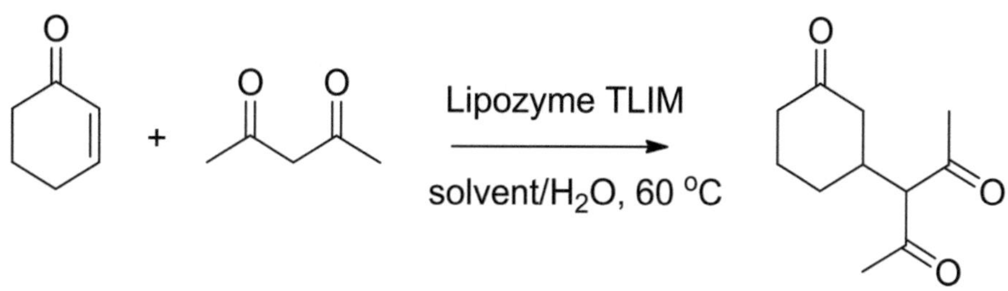

Figure 18.
Michael reaction of cyclohexenone and acetylacetone catalyzed by lipozyme TLIM in different solvents.

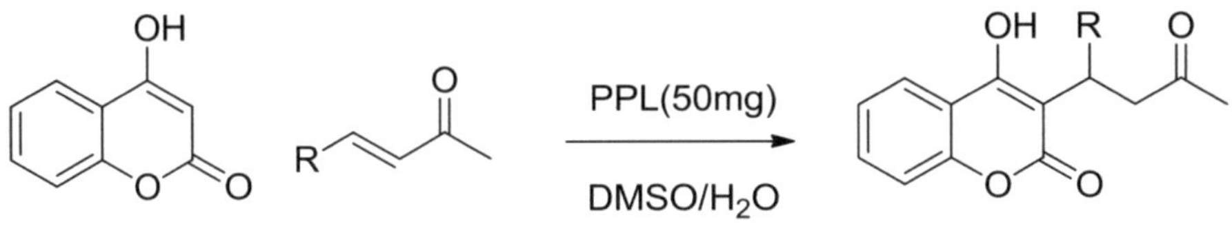

Figure 19.
The Michael addition of 4-hydroxycoumarin 1 to α,β-unsaturated enones for the formation of warfarin and derivatives.

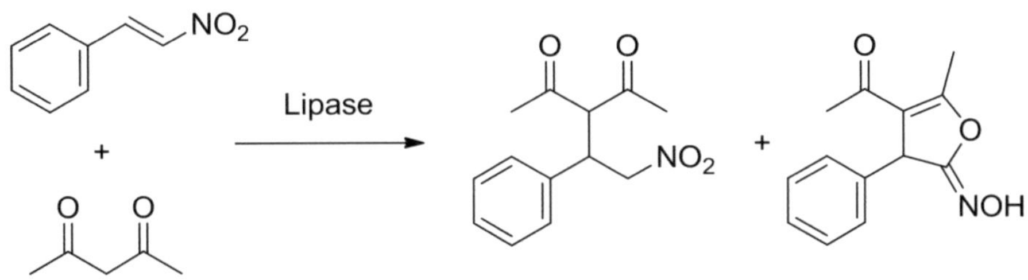

Figure 20.
Lipase-catalyzed reaction of nitrostyrene and acetylacetone.

(**Figure 20**). With the aim to get cyclic product in more efficient manner, the catalytic activities of several lipases were firstly tested in mixed ethanol/water solvents. Among the tested lipases, PPL showed the best activity. And according to the single-crystal X-ray diffraction analysis of cyclic products, the reaction was confirmed to give the product oximes with Z-stereoselectivity.

Then, our group reported for the first time lipase-catalyzed direct vinylogous Michael addition reactions of vinyl malononitriles to nitroalkenes (**Figure 21**) [33]. A series of nitroalkenes reacted with vinyl malononitriles to produce the corresponding products with moderate to high yields in the presence of Lipozyme® (immobilized lipase from *Mucor miehei*). The excellent diastereoselective products were produced in all reactions in acetonitrile at 30°C for 48 h. The enzyme has only a very slight loss of catalyst efficiency after being reused for seven consecutive cycles of the reaction in the previously determined optimized conditions.

The reaction mechanism was studied by computational simulation approach using dock. Based on the proposed catalytic mechanism of Michael reaction, two docking process of the substrates with the amino acids of the active site were performed. The calculation results explained the experimental results that the lipase possessed specific substrate selectivity. To further elucidate the different catalytic effects of RML, CALB, and CRL, structural characteristics of their active site were analyzed, respectively (**Figure 22**). The docking results showed that once vinyl malononitrile was occupied by nitrostyrene, it could not be docked with the active site of CRL. It indicates that the active site is too narrow to bind both two substrates at the same time, so it could not catalyze the direct Michael addition reaction. As for CALB, the active site seems big enough for both nitrostyrene and vinyl malononitrile, but the docking results showed that nitrostyrene blocked proton

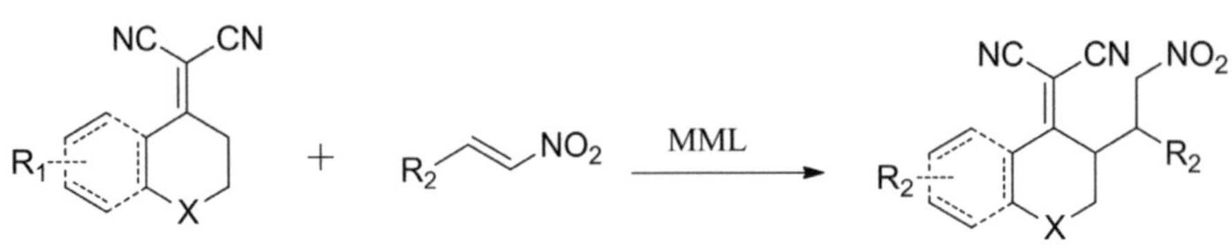

Figure 21.
Lipase-catalyzed direct vinylogous Michael addition reaction.

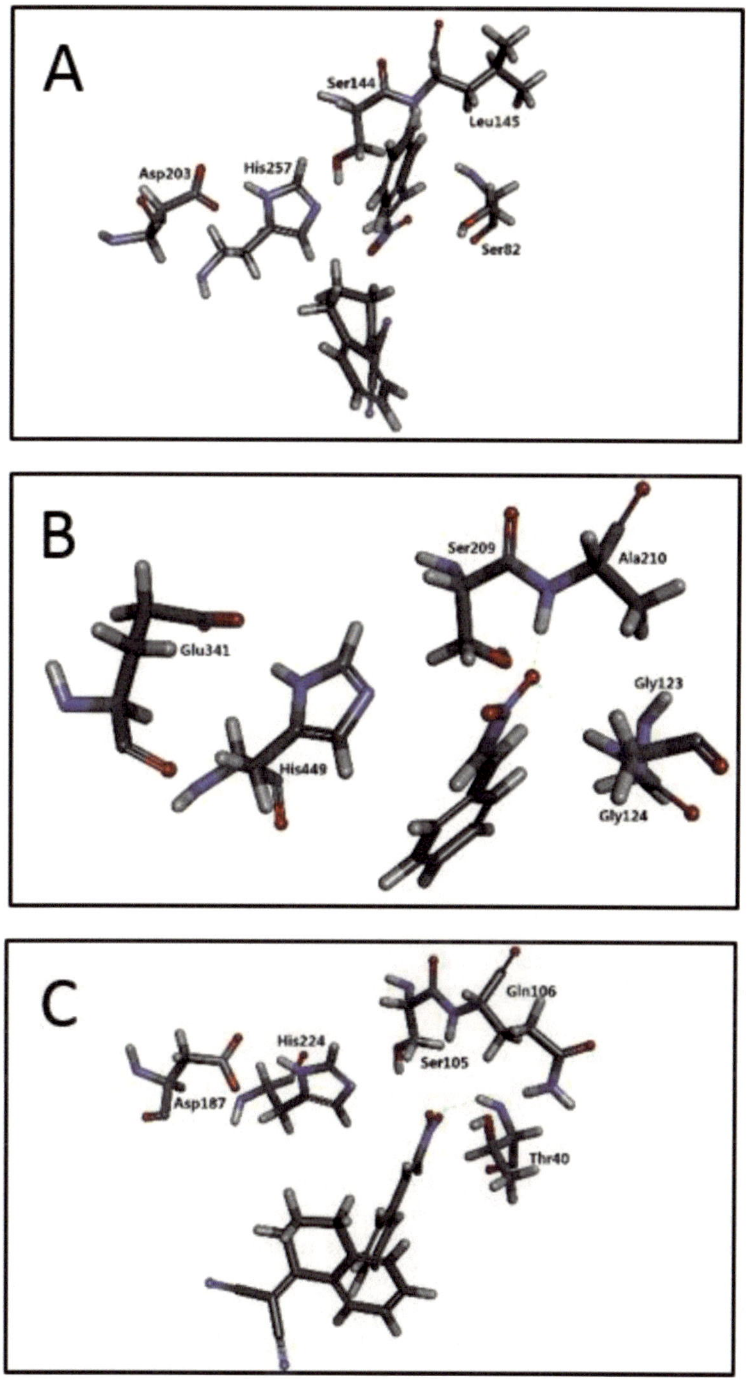

Figure 22.
The comparison of RML, CRL, and CALB: (A) the docking result of RML and substrates, (B) the docking result of CRL and substrates and (C) the docking result of CALB and substrates.

transferring from vinyl malononitrile to histidine, which may make it unable to catalyze the direct vinylogous Michael addition.

Reaction mechanism of the lipase-catalyzed direct vinylogous Michael addition reaction has been proposed (**Figure 23**). First, nitroalkenes bind to oxygen anion pores and were stabilized by three hydrogen bonds with Leu145 and Ser82.

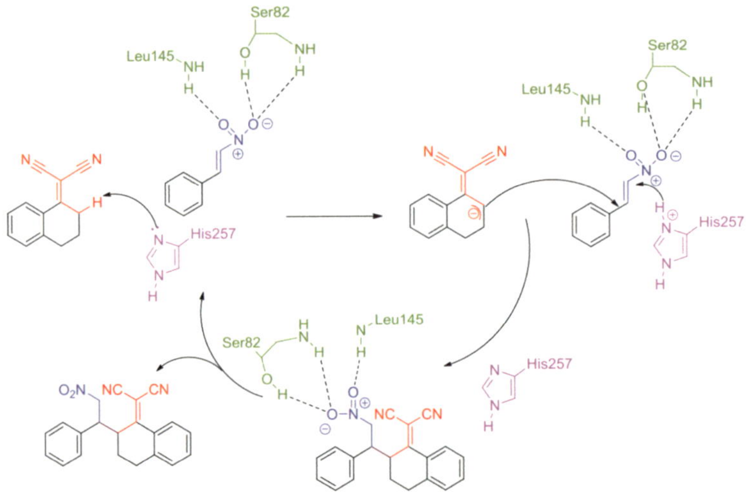

Figure 23.
The proposed mechanism of lipase-catalyzed vinylogous Michael addition.

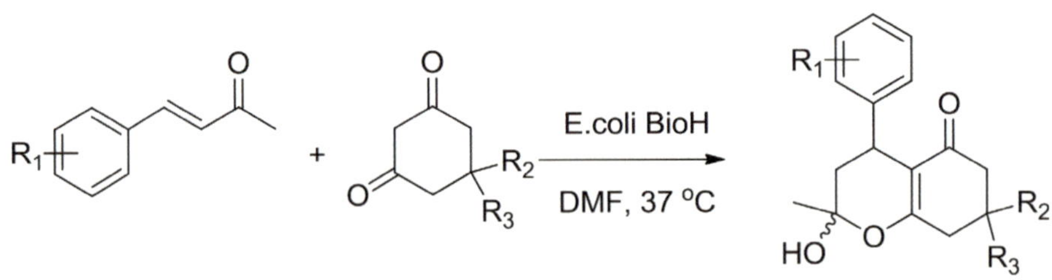

Figure 24.
BioH esterase-catalyzed Michael addition-cyclization cascade reaction.

The protons were then transferred from vinyl malononitrile to His 257 to form a transition state. Subsequently, the protons were transferred from the imidazole group of His 257 to nitroolefins, and the carbon of nitroolefins were attacked by nucleophilic carbon molecules to form products.

In 2014, Ye et al. reported the preparation of 2-hydroxy-2-methyl-4-(4-nitrophenyl)-3,4,7,8-tetrahydro-2H-chromen-5(6H)-one by Michael addition-cyclization cascade reaction of p-nitrobenzalacetone with 1,3-cyclohexanedione in anhydrous media, and control experiments were conducted (**Figure 24**) [34]. The high yield was observed with *Escherichia coli* BioH esterase in DMF at 37°C. In order to preliminarily explore the mechanism of the reaction, site-directed mutagenesis was performed on the hydrolysis catalytic triad of BioH, and the results indicated "alternate-site enzyme promiscuity." Using a series of substituted phenylacetone and 1,3-cyclodiketone as reactants, the yield could reach 76.3%.

3. Aldol reaction

The aldol reaction has long been recognized as one of the most useful tools for organic chemists. The ability to form carbon-carbon bonds can generate a broad range of both natural and novel poly-hydroxylated compounds. Thus, it is the most important and valuable reaction for the preparation of pharmaceuticals, fine chemicals, and natural products. Aldolases have evolved to catalyze the metabolism and catabolism of highly oxygenated metabolites and are found in many biosynthetic pathways of carbohydrates, keto acids, and some amino acids [35]. Aldolases bind their respective donor substrates with high specificity and generally will not accept any other donors, even if their structures are similar to the natural donor. The advantages of using aldolases are very high stereospecificity and environmentally benign reaction conditions [36]. However, the limited number of substrates as well as the high cost of these biocatalysts has led researchers to consider other more stable enzymes [37].

In 2003, Berglund and co-workers firstly reported the serine hydrolase *Candida antarctica* lipase B to have catalytic activity for aldol reactions [38]. Our group reported the first lipase-catalyzed asymmetric aldol reaction in 2008 [39]. However, these aldol reactions in earlier studies involving hydrolases just showed moderate activities and selectivities; some more efficient promiscuous aldol reaction have been researched and presented in the last decade.

3.1 Aldol reaction

In 2012, Guan et al. firstly demonstrated that lipase from porcine pancreas, type II (PPL II), has been observed to catalyze the direct asymmetric aldol reaction of heterocyclic ketones with aromatic aldehydes at 30°C in CH_3CN/H_2O (**Figure 25**) [40]. PPL II has good catalytic activity and good adaptability to different substrates. Its enantioselectivity can reach 87% ee and enantioselectivity 83:17 (anti/syn). Then PPL II has aldolase function in organic solvents.

In the same year, the same group also reported the similar asymmetric aldol reaction of aromatic and heteroaromatic aldehydes with cyclic and acyclic ketones in acetonitrile in the presence of a phosphate buffer by chymopapain, which is a cysteine proteinase isolated from the latex of the unripe fruits of *Carica papaya* [41]. Chymopapain exhibited the best catalytic activity and moderate stereoselectivity in DMSO, and the enzyme showed the best enantioselectivity of 79% ee in CH_2Cl_2 with low diastereoselectivity (**Figure 26**). In consideration of both diastereo- and enantioselectivities, the group chose MeCN as a suitable solvent for the asymmetric direct aldol reaction, which gave the best dr of 77:23 and a moderate ee of 76% among the tested solvents. Then, in order to further optimize the direct asymmetric aldol reaction catalyzed by papain, the effects of water content,

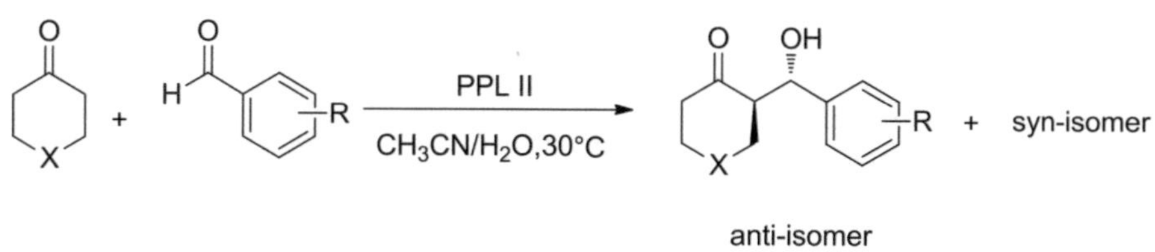

Figure 25.
Lipase-catalyzed direct asymmetric aldol reaction of heterocyclic ketones with aromatic aldehydes.

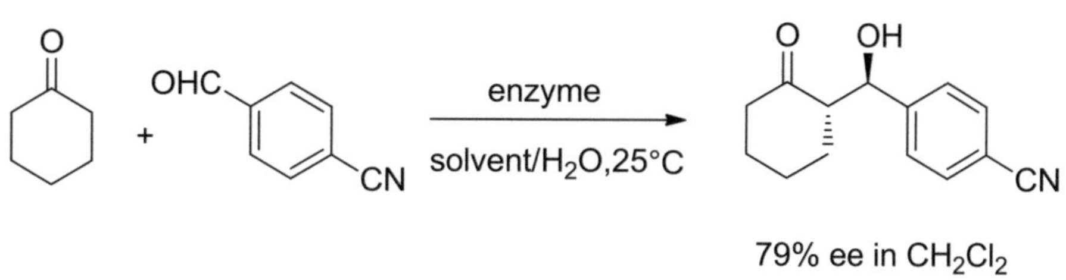

Figure 26.
The asymmetric aldol reaction of 4-cyanobenzaldehyde and cyclohexanone.

reaction temperature, and the amount of buffer on the enzymatic reaction were investigated. The reaction of 4-cyanobenzaldehyde with cyclohexanone was used as a model reaction.

The authors proposed a mechanism for the chymopapain-catalyzed aldol reaction (**Figure 27**). The catalytic triad of Cys, His, and Asn formed the active site of chymopapain. Firstly, the carbonyl of the substrate ketone is coordinated in the oxygen anion pore of Asn-His binary and active center. Secondly, a proton is transferred from the ketone to the His residue to form enolate ion. Thirdly, another substrate aldehyde accepts the proton from imidazolium cation and forms a new carbon-carbon bond with ketones. Finally, the product is released from the oxyanion hole and separates from the active site.

In 2013, our group firstly reported the asymmetric aldol reaction between aromatic aldehydes and cyclic ketones by PPL (**Figure 28**) [42]. The results showed that a small amount of water could promote the catalytic activity of PPL at 37°C. A wide range of aromatic aldehydes reacted with cyclic ketones to provide the corresponding aldol products with high yields (up to 99%) and moderate to good stereoselectivity (up to 90% ee and 99:1 dr).

In the same year, a simple and convenient synthesis route of series α,β-unsaturated aldehydes was formed by combining the two catalytic activities of the same enzyme with the one-pot method of aldehyde-alcohol reaction and in situ acetaldehyde formation (**Figure 29**) [43]. Lipase from *Mucor miehei* has conventional and promiscuous catalytic activities for the hydrolysis of vinyl acetate and aldol condensation with in situ-formed acetaldehyde.

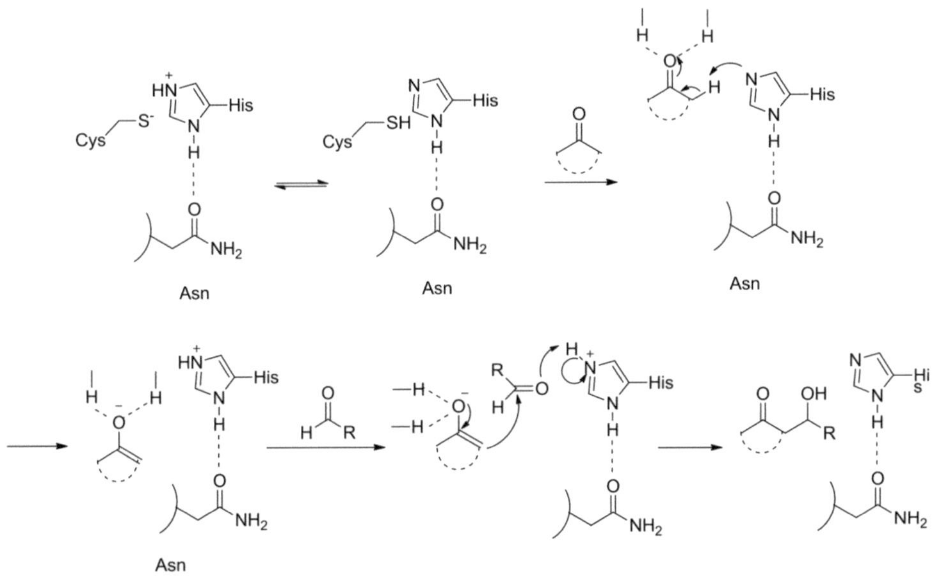

Figure 27.
Proposed mechanism for the chymopapain-catalyzed aldol reaction.

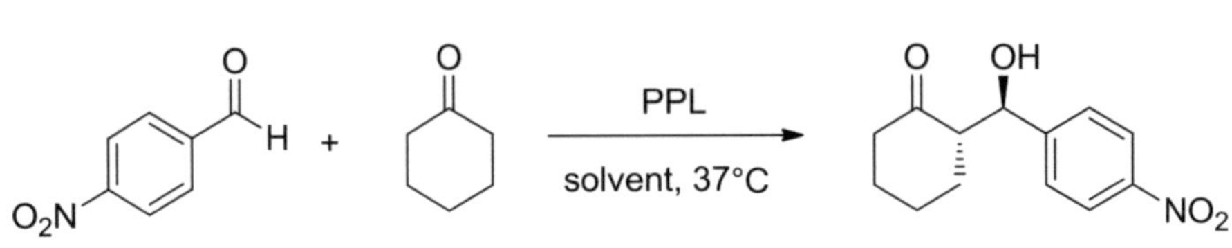

Figure 28.
The asymmetric aldol reaction between aromatic aldehydes and cyclic ketones by PPL.

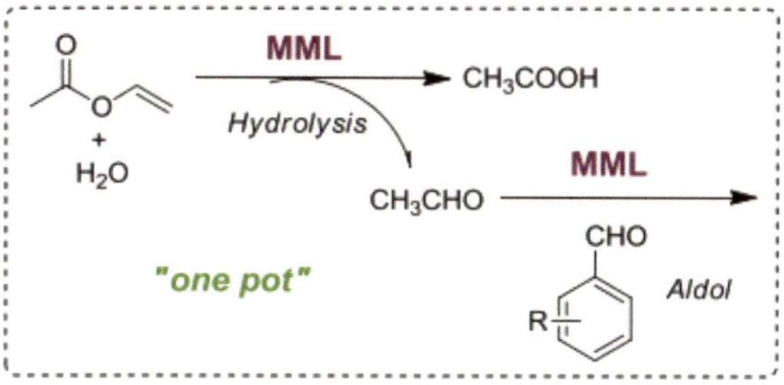

Figure 29.
MML-catalyzed aldol condensation using an in situ-generated acetaldehyde.

In 2014, Majumder and Gupta found that the properties of lipase-catalyzed reaction products of acetylacetone with 4-nitrobenzaldehyde depend on the source of lipase and reaction medium (**Figure 30**) [44]. *Mucor javanicus* lipase was found to give 70% aldol and 80% enantiomeric excess in anhydrous t-amyl alcohol.

Gao and Guo et al. demonstrated the catalytic promiscuity of an acyl-peptide releasing enzyme from *Sulfolobus tokodaii* (ST0779) for aldol addition reaction for the first time, and accelerated activity was observed at elevated temperature (**Figure 31**) [45]. The turnover number kcat (s^{-1}) of this thermostable enzyme at 55°C is 7.78-fold higher than that of PPL at its optimum temperature of 37°C, and the molecular catalytic efficiency kcat/Km (M^{-1} s^{-1}) adds up to 140 times higher than PPL.

The authors proposed a mechanism for the ST0779-catalyzed aldol reaction between acetone and 4-nitrobenzaldehyde (**Figure 32**). Based on the structure simulation of ST0779, the aldol reaction catalyzed by ST0779 with acetone and p-nitrobenzaldehyde as model reaction was proposed. Because of its thermodynamic superiority and high affinity, acetone first enters the active site and then is accommodated by the active site residues Ser439 and His 555. Proton transfer forms a transition state of enol salts, which is stable by Ser439. Asp523 is involved in stabilizing the positive charge of His 555-protonated imidazole ring. In the next

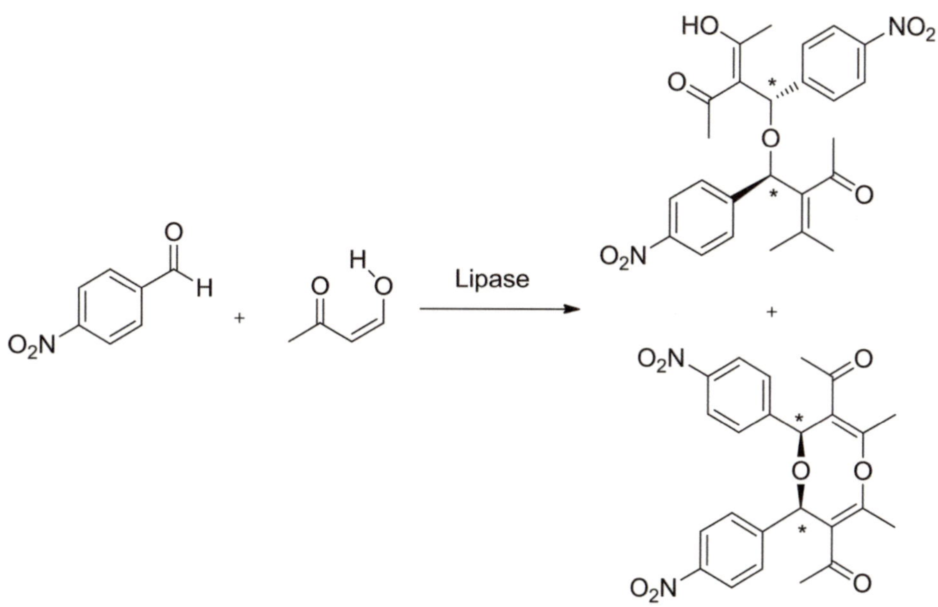

Figure 30.
The lipase-catalyzed promiscuous reaction: formation of acyclic and cyclic 2:2 adducts.

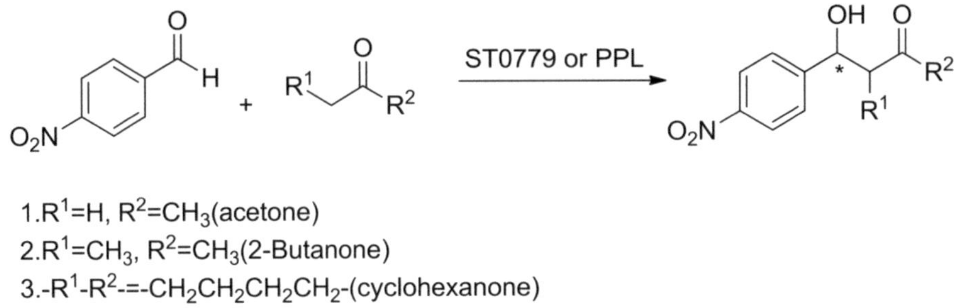

Figure 31.
Aldol reactions catalyzed by ST0779 or PPL.

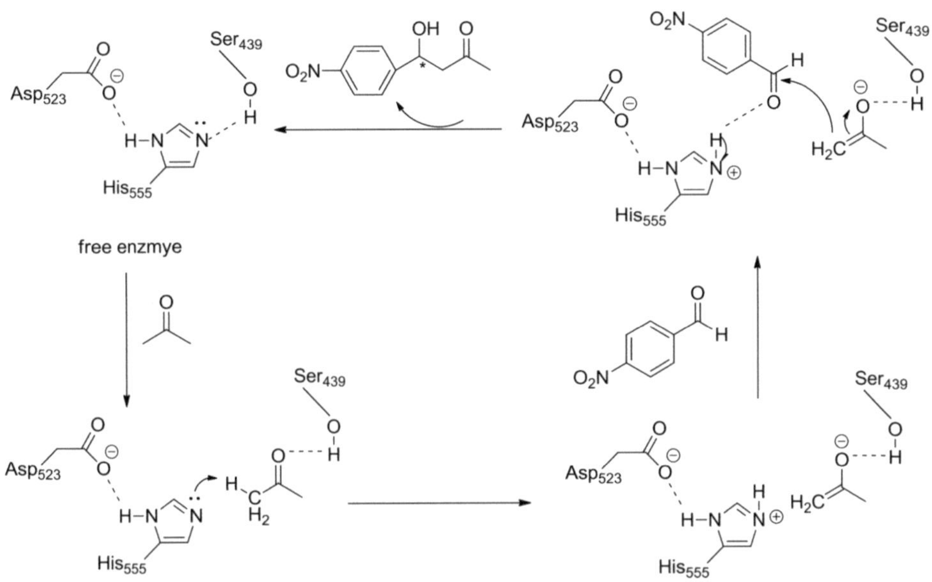

Figure 32.
Proposed mechanism for ST0779-catalyzed aldol reaction between acetone and 4-nitrobenzaldehyde.

step, the oxygen of carbonyl group in 4-nitrobenzaldehyde is protonated by protons from His 555 imidazole ring, and the carbon atoms in the same carbonyl group are neutrally attacked by oleic acid carbon to form a new C—C bond. Finally the aldol product is released from the enzyme, and the enzyme is freed for a new reaction.

In 2016, Wu et al. demonstrated a one-pot bienzymatic cascade in organic media to synthesize chiral β-hydroxy ketones for the first time (**Figure 33**) [46]. The decarboxylative aldol reaction catalyzed by an immobilized lipase from *Mucor miehei* (MML) and the synthesis of β-hydroxy ketone catalyzed by a lipase A or B from *Candida antarctica* (CALA or CALB) were combined in this one-pot protocol, reducing the purification step between the two reactions. (S)-β-hydroxy ketones and acylated (R)-β-hydroxy ketones could be obtained under mild reaction condition. The ee values of most chiral compounds were in a range of 94–99%, while the total yields of both chiral products were all above 85%. This enzymatic one-pot chain method is still very effective, not only can it be amplified to the level of grams but also the catalyst was recovered three times.

In 2019, Gao et al. demonstrated the construction of an unencapsulated remote-controlled nanobiocatalytic system. The system used three enzyme-conjugated gold nanorod composites (EGCs) to control reaction rates in real time by self-assembling

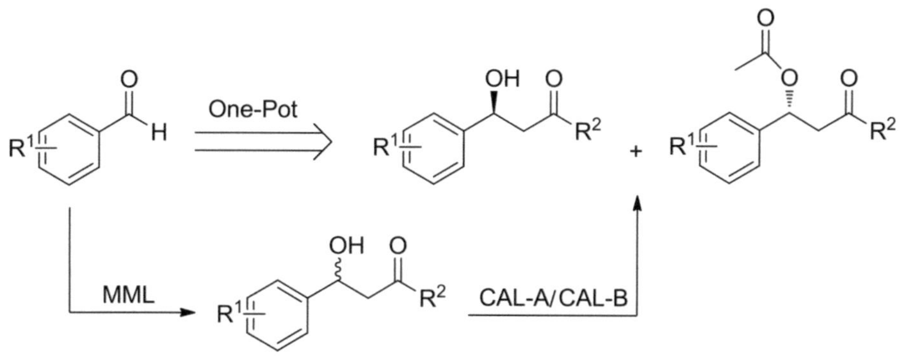

Figure 33.
One-pot cascade for synthesis of chiral β-hydroxy ketone derivatives.

enzymes formed by the combination of enzymes at different optimal temperatures and gold nanorods (GNRs) [47]. By using the photothermal effect of GRS to transfer energy quickly, coupled with the real-time and long-range regulation of enzyme activity, improving the thermal stability of the enzyme and effective catalysis of the aldol reaction can be achieved. The increase in energy inside GRS, stimulated by distant near-infrared (NIR) stimuli, leads to increased enzyme activity. The results show that the method of internal heating that transfers energy more directly to the enzyme-catalyzed site is a faster and more effective energy transfer method. The results also show that the catalytic effect of the remote-controlled nanocatalytic system at lower temperature is the same as that of the free enzyme at higher temperature, but it has the advantages of improving the stability of the enzyme and extending its service life. Specifically, PPL EGCs at room temperature exhibit the same catalytic effect as achieved by free PPL at 40°C, while ST0779EGCs and APE1547 EGCs at 33°C exhibit a higher catalytic effect than their corresponding free enzymes at 63°C. In addition, EGCs have superior catalytic efficiency and product yield compared with aldol addition in free enzyme systems.

3.2 Henry (nitroaldol) reaction

The nitroaldol or Henry reaction is one of the most useful carbon-carbon bond-forming reactions and has wide synthetic applications in organic chemistry. This reaction provides access to valuable racemic and optically active β-nitro alcohols, which are very useful in organic synthesis as precursors for pharmaceutical and biological purposes. In recent years, efficient nonconventional biocatalytic approaches have been reported [48].

Guan et al. firstly reported transglutaminase was used to catalyze Henry reactions of aliphatic, aromatic, and heteroaromatic aldehydes with nitroalkanes (**Figure 34**) [49]. The reactions were carried out at room temperature, and the corresponding nitroalcohols were obtained in yields up to 96%.

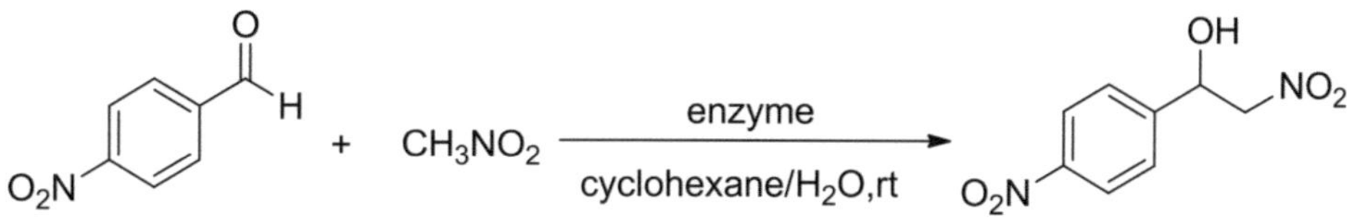

Figure 34.
Enzyme-catalyzed Henry reaction of 4-nitrobenzaldehyde and nitromethane.

Then, the same group reported glucoamylase from *Aspergillus niger* (AnGA) catalyze Henry reactions of aromatic aldehydes and nitroalkanes in 2013 [50]. The reactions were carried out at 30°C in the mixed solvents of ethanol and water, and the corresponding β-nitro alcohols were obtained in yields of up to 99%. Experiments demonstrated that AnGA could be inhibited by the product of the Henry reaction at 80°C. This enzymatic Henry reaction has a broad substrate scope and could be facilely enlarged to gram scale. Based on the experiments with denatured and inhibited AnGA, and the comparison of natural activity and promiscuous activity, the possible mechanism was also discussed (**Figure 35**). Glu400, as a base, deprotonates the α-carbon of the nitroalkane providing intermediate I. At the same time, Glu179, as an acid, donates a proton to the carbonyl oxygen of the aldehyde generating intermediate II. Then, the a-carbon of I, as a nucleophile, attacks the carbonyl of II forming a new carbon-carbon bond. Finally, the product (β-nitro alcohol) is released from the active site.

On the other hand, Lin and co-workers demonstrated the Henry reaction can also be catalyzed in a neat organic solvent. When using the D-aminoacylase from *Escherichia coli* as the promiscuous biocatalyst, DMSO was found to be the best solvent at 50°C [51]. Interestingly, the synthesis of optically active β-nitro alcohols was achieved by a two-step strategy combining the D-aminoacylase-catalyzed nitroaldol reaction with the PSL-catalyzed resolution of the so obtained racemic β-nitro alcohols (**Figure 36**) [52]. Both alcohols and acetates were isolated in good yields and high enantiomeric excess (>84% ee_s; >96% ee_p; E > 150).

In 2013, lipase A from *Aspergillus niger* was used in the Henry reaction between aromatic aldehydes and a large excess of nitroalkanes in an organic/water medium (**Figure 37**) [53]. The yield of corresponding β-nitro alcohols at 30°C reached 94%.

Gotor and co-workers reported the inexpensive carrier protein bovine serum albumin (BSA) as catalyst was firstly used in the condensation of an appropriate aldehyde with 1-nitroalkanes in aqueous media (**Figure 38**) [54]. By optimizing the reaction conditions, the yield of corresponding nitroalcohols at 30°C reached 91%.

Similarly, two other well-known lipases, *Pseudomonas cepacia* lipase and CALB, were found to catalyze the Henry reaction [55]. Nevertheless, spectroscopic experiments showed that the immobilization protocols contribute to the change in the secondary structure of the enzyme, which leads to improved conversion rates.

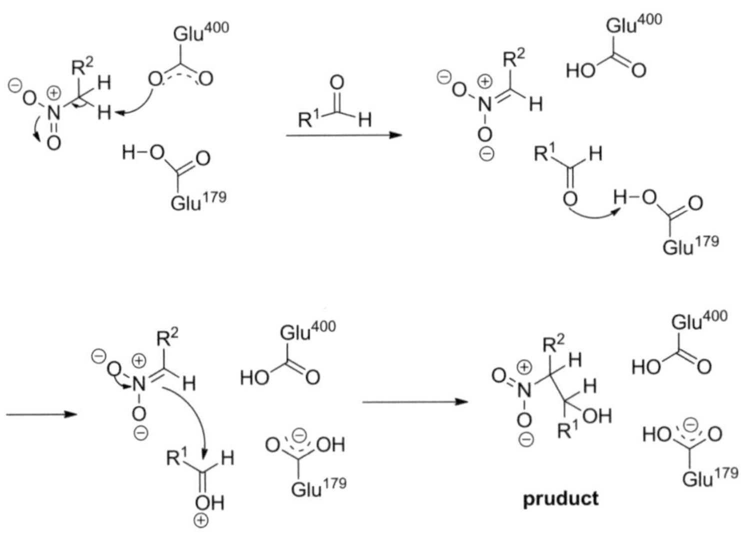

Figure 35.
Possible mechanism of the AnGA-catalyzed Henry reaction.

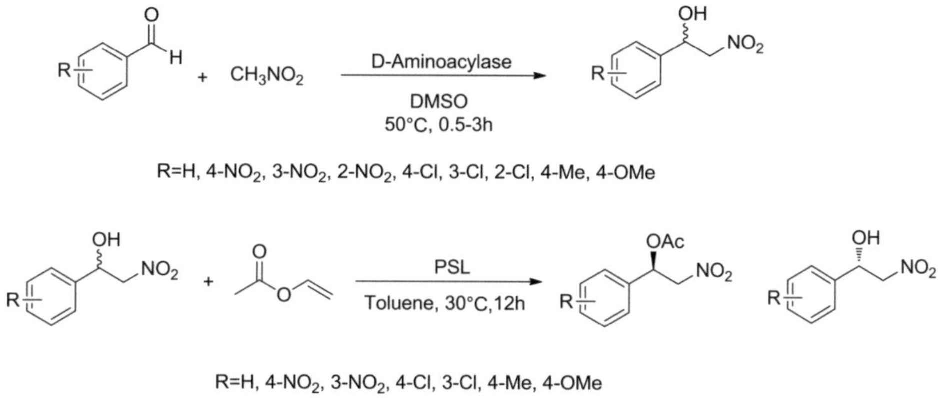

Figure 36.
Two-step method to obtain β-nitro alcohols and the corresponding acetates of both configurations based on a D-aminoacylase-catalyzed reaction and PSL-mediated kinetic resolution using vinyl acetate as acyl donor.

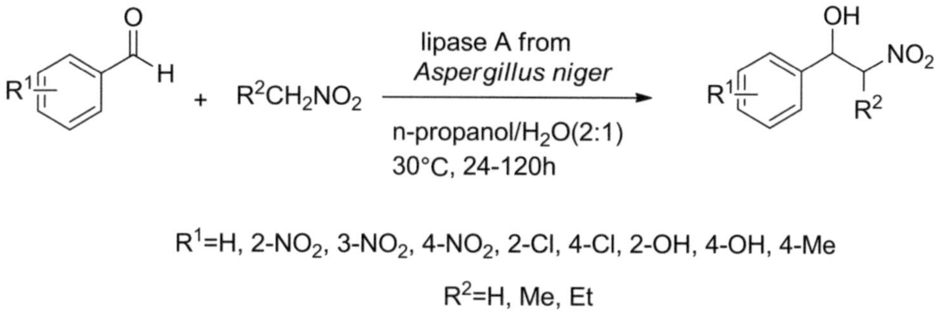

Figure 37.
Nitroaldol reaction between aromatic aldehydes and nitroalkanes catalyzed by lipase A from Aspergillus niger.

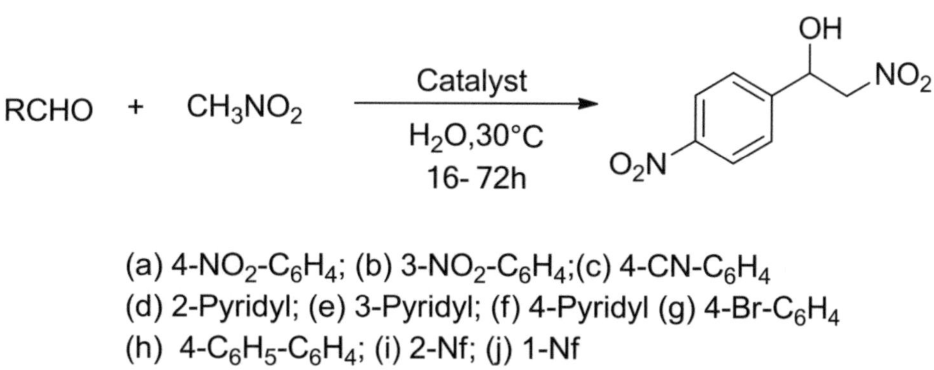

Figure 38.
Catalytic nitroaldol addition between different aromatic aldehydes and nitromethane.

3.3 Aldol (nitroaldol) reaction in untraditional solvent

Solvents for a biocatalysis reaction have experienced several generations of development. Traditional organic solvents (water miscible or water immiscible), in the form of cosolvents or second phase, can provide solutions for the above-described challenges. However, organic solvents inevitably face their own disadvantages, such as high volatility, difficulty in preparation, and inhibition of the activity of biocatalysts. So the untraditional solvents such as buffer solvent, ionic liquids (ILs), and deep eutectic solvents (DESs) have attracted the interest of many groups.

In 2013, our group demonstrated bovine pancreatic lipase (BPL) was first used to catalyze the aldol reaction and acidic buffer was first used for promiscuous enzymatic aldol reaction (**Figure 39**) [56]. The highest yield (99.0%), the best dr of 96:4, and a moderate ee of 66% were observed with aromatic aldehyde and ketone by BPL in phosphate-citrate buffer (pH 5.6, 5.0 mL) at 30°C.

Porto et al. demonstrated the lipase from *Rhizopus niveus* (RNL) catalyzed by unspecific protein catalysis the aldol reactions between cyclohexanone and aromatic aldehydes in organic solvents with water or aqueous buffer solution (**Figure 40**) [57]. The reactional conditions strongly influenced the yield (0–99%) and enantioselectivities in the anti-products (6–55% ee). The aldol products with enantioselectivities in the anti-product were observed for inactive enzyme and in denaturing conditions. Therefore, the reactions in the evaluated conditions were proceeded by unspecific protein catalysis with moderate enantioselectivities and not by promiscuous activity.

Ionic liquids are the first enzyme-compatible untraditional media developed by the green and sustainable concept (given their low vapor pressure). Numerous reactions, e.g., hydrolytic and redox reactions as well as formation of C—C bond, have been successfully performed in such ILs-containing media. We demonstrated PPL was used to catalyze asymmetric cross aldol reactions of aromatic and heteroaromatic aldehydes with various ketones in ionic liquid ([BMIM][PF_6]) for the first time in 2014 (**Figure 41**) [58]. PPL exhibited high catalytic activity and

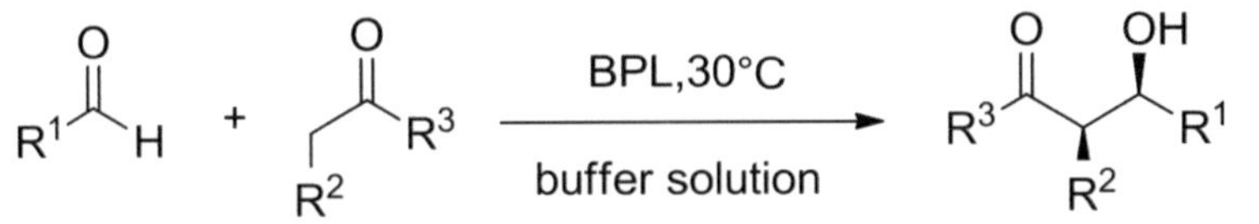

Figure 39.
The BPL-catalyzed asymmetric aldol reaction in buffer solution.

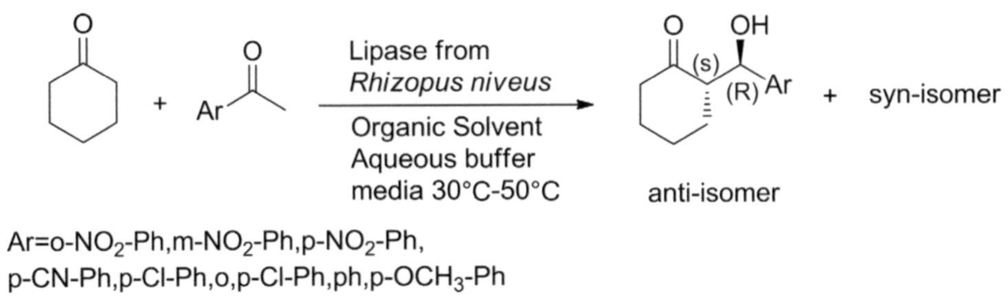

Figure 40.
Aldol reactions by lipase from Rhizopus niveus.

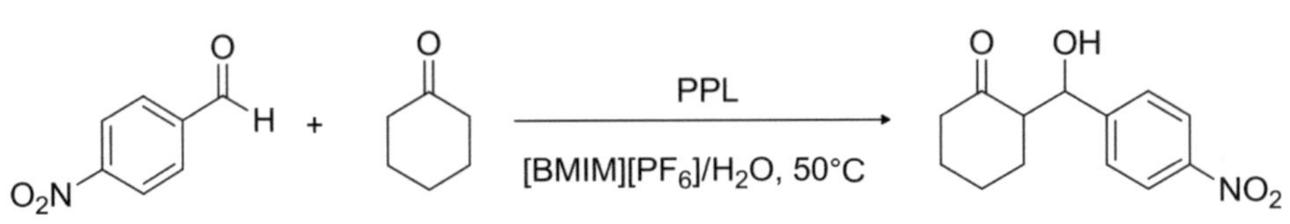

Figure 41.
The PPL-catalyzed asymmetric cross aldol reaction in ionic liquid.

excellent stereoselectivity in this efficient and recyclable room-temperature ionic liquid in the presence of moderate water. High yields of up to 99%, excellent enantioselectivities of up to 90% ee, and good diastereoselectivities of up to >99:1 dr were achieved.

Despite the excellent performance of ILs in biocatalysis, more doubts about their ungreenness and environmental influence have been gradually presented. Deep eutectic solvents, the recognized alternative of ILs, first came to the public vision in 2001. Since then, research on DESs faced a prosperous increase in many fields, such as extraction, materials synthesis and biotransformation, and biocatalysis.

Gotor-Fernández and co-workers reported a promiscuous lipase-catalyzed aldol reaction has been performed for the first time in DESs in 2016. The aldol reaction between 4-nitrobenzaldehyde and acetone was examined in-depth, with excellent compatibility being found between PPL and DESs (choline chloride/glycerol mixtures) for the formation of the aldol product in high yields (**Figure 42**) [59]. The system was compatible with a series of aromatic aldehydes and ketones including acetone, cyclopentanone, and cyclohexanone.

At the same year, Tian et al. explored the Henry reaction catalyzed by lipase AS using deep eutectic solvents as a reaction medium (**Figure 43**) [60]. The studies had shown that adding 30 vol% water to DES could increase the catalytic activity of enzymes. The final yield of the lipase AS-catalyzed Henry reaction was 92.2% in a DES-water mixture within only 4 h. In addition, the lipase AS activity was improved by approximately threefold in a DES-water mixture compared with that in pure water. The methodology was also extended to the aza-Henry reaction. The enantioselectivity of both Henry and aza-Henry reactions was not found.

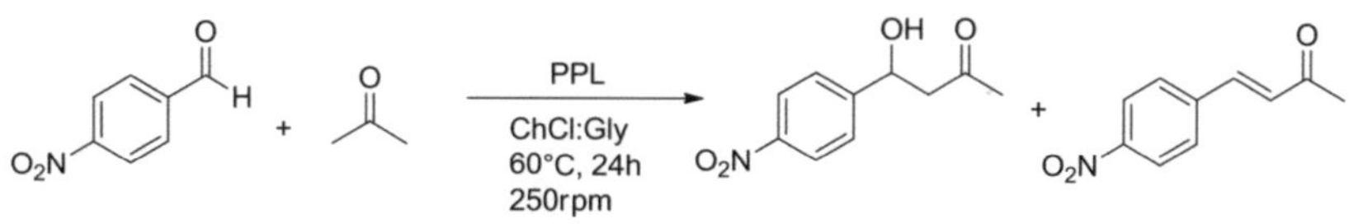

Figure 42.
The PPL-catalyzed asymmetric cross aldol reaction in DESs.

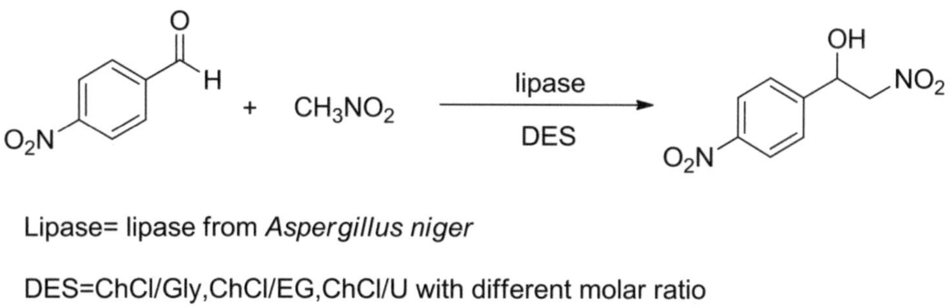

Figure 43.
Lipase AS-catalyzed Henry reaction in DES.

4. Multicomponent reactions (MCRs)

Multicomponent reactions have attracted sustained attention because they represent a powerful tool for the construction of complex molecular structures with evident advantages, such as simplified workup procedures, high overall yields, and versatile product libraries. Recently, hydrolases have allowed the development of multiple transformations and mainly served for the synthesis of heterocyclic compounds with high complexity in high yields. In this section, we will focus on the hydrolase-catalyzed multicomponent reaction in a one-pot transformation.

4.1 Mannich reaction

The Mannich reaction is a typical and the first example for the hydrolase-catalyzed multicomponent reaction, which is atom-economic and a powerful synthetic method for generating carbon-carbon bonds and nitrogenous compounds. An unprecedented "one-pot," direct Mannich reaction of ketone, aldehyde, and amine catalyzed by lipase was described first in 2009 [61]. Lipase from *Mucor miehei* (MML) efficiently catalyzed the Mannich reaction (**Figure 44**).

To assess the generality of the lipase-catalyzed Mannich reaction, we extended other substrates such as cyclohexanone, butanone, and 1-hydroxy-2-propanone in more benign reaction system (ethanol/water) catalyzed by the lipase from *Candida rugosa* (CRL) [62]. It was found that a wide range of aromatic aldehydes could effectively participate in the CRL-catalyzed Mannich reaction to give the corresponding β-amino carbonyl compounds (**Figure 45**). The reaction was favored by the electron-withdrawing substituents of the aldehydes.

In 2012, Guan et al. reported the enzyme-catalyzed, direct, three-component asymmetric Mannich reaction using protease type XIV from *Streptomyces griseus* (SGP) in acetonitrile (**Figure 46**) [63]. This characteristic makes it important to develop an enzyme-catalyzed asymmetric Mannich reaction as a more sustainable

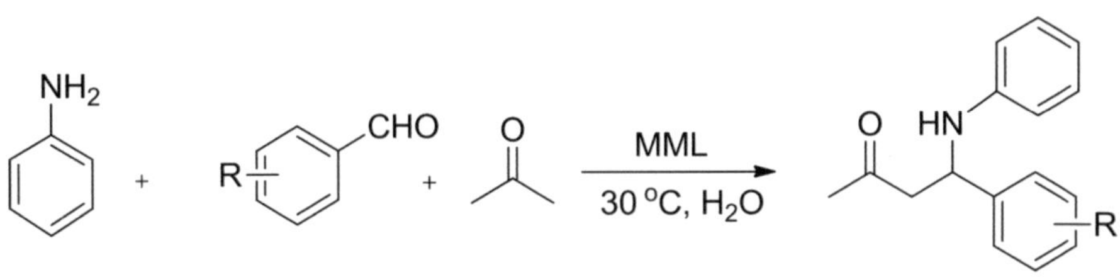

Figure 44.
The first lipase-catalyzed direct Mannich reaction.

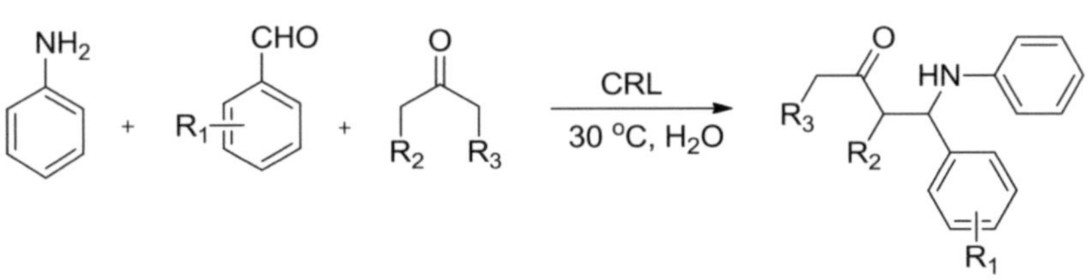

Figure 45.
Lipase-catalyzed direct Mannich reaction of various aryl aldehydes and ketones with aniline.

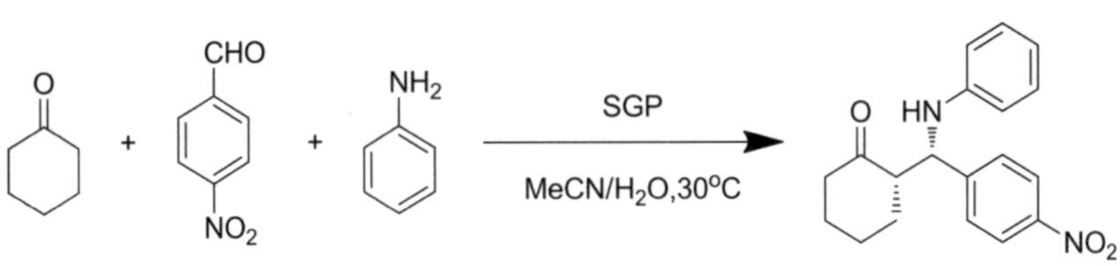

Figure 46.
SPG-catalyzed direct Mannich reaction.

complement to chemical catalysis. The control experiments with the denatured enzyme and non-enzyme proteins indicated that the specific natural fold of SGP was responsible for its stereoselectivity in the Mannich reaction. A wide range of substrates were accepted by the enzyme, and yields of up to 92%, enantioselectivities of up to 88% ee, and diastereoselectivities of up to 92:8 dr were achieved. As an example of enzyme catalytic promiscuity, this work broadens the scope of SGP-catalyzed transformations.

In the same year, Lin et al. inspired by chemical cofactors or mediators expect some small molecules to similarly improve the enzymatic Michael addition of unactivated carbon nucleophiles [64]. They found that the CALB/acetamide co-catalyst system can effectively catalyze the Michael addition between less-activated ketones and aromatic nitroolefins. This is of particular interest because neither CALB nor acetamide can independently catalyze the reaction to any significant extent. The CALB/acetamide catalyst system is also effective for other C—C bond-forming reactions with varying degrees of success, for example, CALB-catalyzed Mannich reaction (**Figure 47**). After adding acetamide as a co-catalyst, the yield increased by 50% (from 25 to 38%). The synergistic catalytic system of the lipase and the small molecule organic catalyst will greatly expand the application prospect of the enzyme in organic synthesis.

After 2 years, Guan et al. reported the use of acylase I from *Aspergillus melleus* in the asymmetric Mannich reaction (**Figure 48**) [65]. Compared to the current chemical technologies, this enzymatic reaction is more environmentally friendly and sustainable by using biocatalysts from inexpensive renewable resources. The activity and stereoselectivity of AMA can be improved by adjusting the solvent, pH, water content, temperature, substrate molar ratio, and enzyme loading. A wide range of substrates can be accepted by AMA, achieving enantioselectivities up to 89% ee, diastereoselectivities up to 90:10 dr, and yields up to 82% in the mixture of MeCN and phosphate buffer pH 8.1 (85:15 v/v 1 mL) at 30°C. This work not only provides new examples of enzyme-catalyzed reliability and potential synthetic methods of organic chemistry but may also help to better understand the metabolic pathways of nitrogen-containing compound biosynthesis.

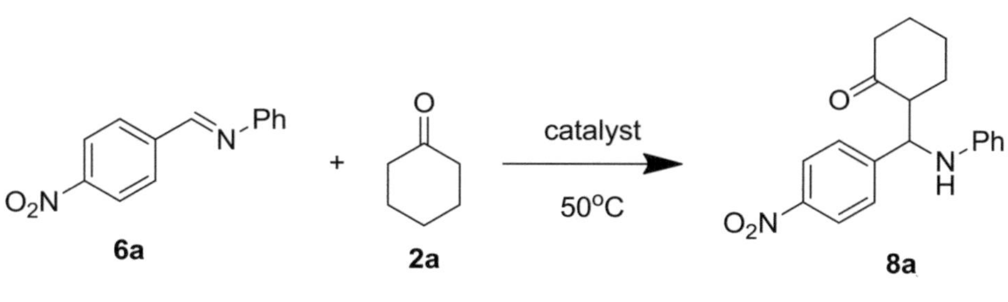

Figure 47.
Mannich reaction of (E)-N-(4-nitro-benzylidene)aniline with cyclohexanone.

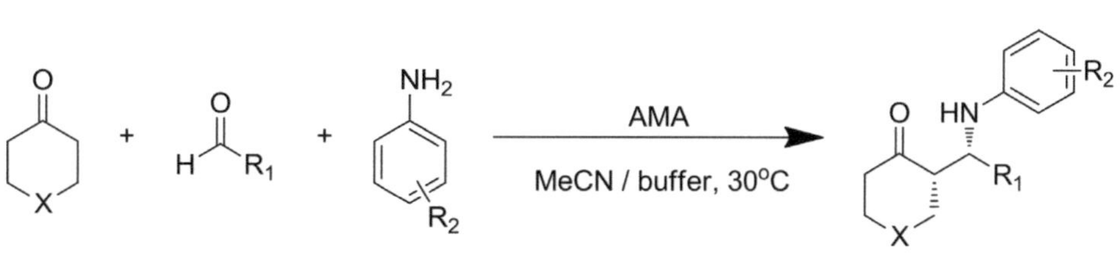

Figure 48.
The AMA-catalyzed Mannich reactions.

4.2 Biginelli reactions

In 2013, Sinha et al. reported bovine serum albumin promoted simple and efficient one-pot procedure for synthesis of 3,4-dihydropyrimidin-2(1H)-ones including potent mitotic kinesin Eg5 inhibitor monastrol under mild reaction conditions (**Figure 49**) [66]. After the reaction conditions are optimized, the yields reached up to 82% in EtOH at 60°C. The catalyst recyclability and gram-scale synthesis have also been demonstrated to enhance the practical utility of process.

Followed by our group, we reported trypsin as a multifunctional catalyst for synthesis of 3,4-dihydropyrimidin-2(1H)-ones by the Biginelli reaction of urea, β-dicarbonyl compounds, and in situ-formed acetaldehyde (**Figure 50**) [67]. This one-pot multistep reaction consists of two relatively independent reactions, both of which are catalyzed by trypsin. First is the transesterification of ethyl acetate and isobutanol at 60°C to produce in situ acetaldehyde, followed by in situ-generated acetaldehyde, urea, and β-dicarbonyl compounds for Biginelli reaction. The first reaction continuously provides a substance for the second reaction, effectively reducing the volatilization loss, oxidation, and polymerization of acetaldehyde and avoiding the negative influence of excess acetaldehyde on the enzyme. Under optimal conditions, a wide range of substrates participate in the reaction and provide the target product in high yield.

In 2017, CALB-catalyzed for synthesis of 3,4-dihydropyrimidin-2(1H)-ones by a tandem multicomponent reaction in one pot (**Figure 51**) has been reported [68]. Several control experiments were performed using acetaldehydes directly to explore the possible mechanism of this procedure. Moreover, owing to the distinct modularity and highly efficient features of the MCR, it assembles libraries of structurally diverse products and provides an exceptional synthesis tool for the discovery of the minimal deep-blue luminogen in the solid state, namely, a single ring. A few of the compounds show deep-blue emissions which only contain a single ring. This is an important application of green biocatalytic promiscuity for constructing a wide variety of new materials.

4.3 Hantzsch reaction

Lin et al. reported a three-component (aldehyde, 1,3-dicarbonyl compound, and acetamide) Hantzsch-type reaction in anhydrous solvent, which gave 1,4-dihydropyridines in moderate to good yields (**Figure 52**) [69]. The group used acetamide as a new source of ammonia. Initially the yield of the reaction with CALB

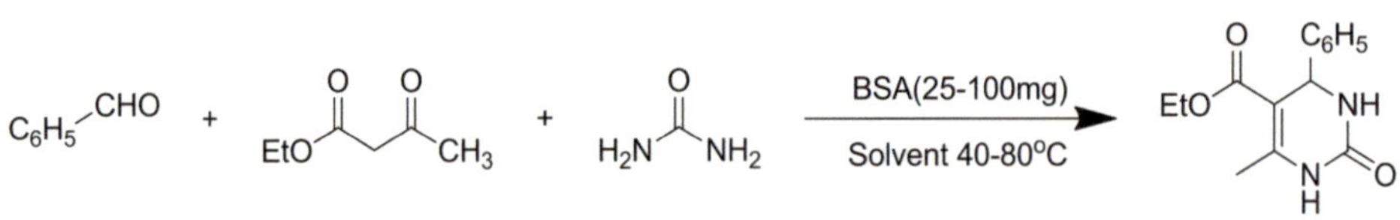

Figure 49.
BSA-catalyzed Biginelli reaction.

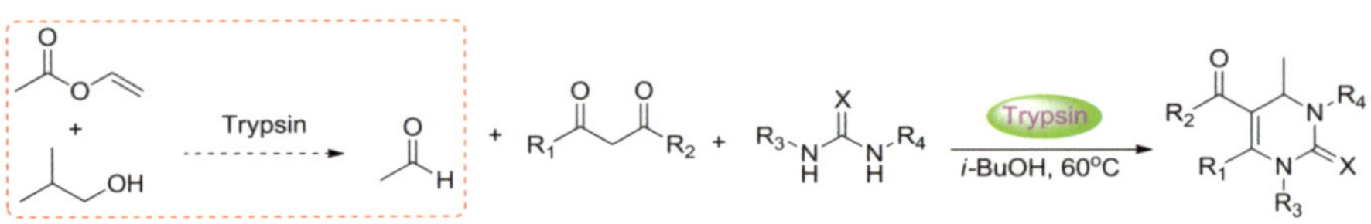

Figure 50.
Trypsin-catalyzed Biginelli reaction using an in situ-generated acetaldehyde.

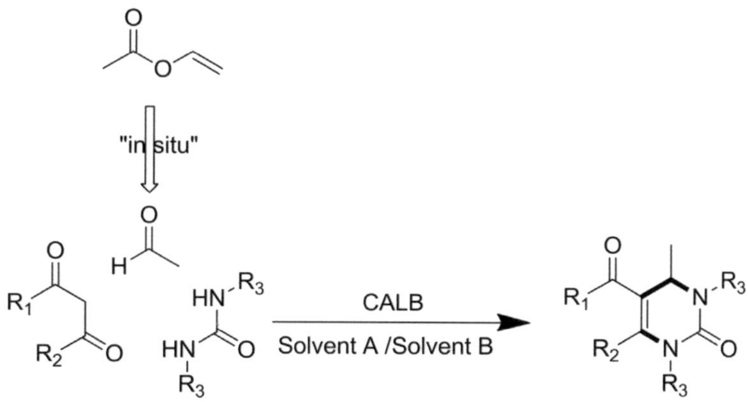

Figure 51.
CALB-initiated tandem Biginelli reaction.

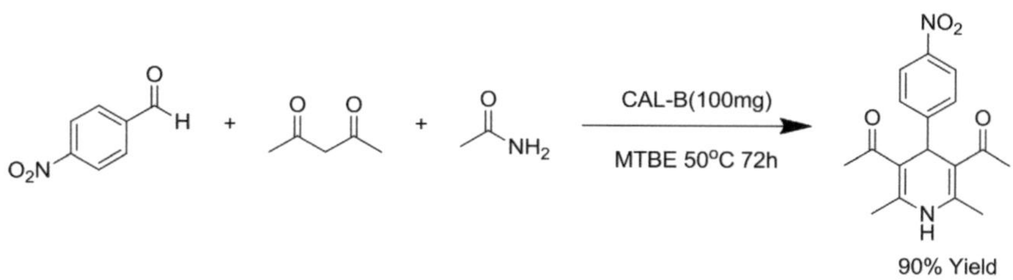

Scheme

Figure 52.
Lipase-catalyzed Hantzch-type reaction.

was very low (only 25%). The yield of the product was slightly improved using a mixed solvent (the ratio of MTBE to acetylacetone was 6:4), and the molar ratio of 4-nitrobenzaldehyde acetamide was 1:4 at 50 mg/ml lipase. When the lipase concentration (100 mg/mL) was increased, the yield increased sharply.

They proposed a reasonable mechanism of the reaction, wherein Asp-His dyad and oxyanion hole in the active site stabilized acetamide (**Figure 53**). This activated acetamide reacted with 1,3-dicarbonyl compounds to form an intermediate, which upon subsequent hydrolysis by CALB formed an enamine intermediate. During this period, a CALB-catalyzed Knoevenagel condensation reaction of the 1,3-dicarbonyl compound with aldehyde formed a separate intermediate (α,β-unsaturated carbonyl compound). Subsequently, the intermediate that is stabilized by the catalytic center of the lipase forms the final product (1,4-dihydropyridine) by Michael addition and intramolecular condensation.

In 2017, our group reported a series of 1,4-dihydropyridines was produced via facile enzymatic Hantzsch reactions in one pot, using acetaldehydes/aromatic aldehydes prepared in situ (**Figure 54**) [70]. After screening several parameters on a model reaction, the tandem process afforded 1a in 80% yield. This approach provided an opportunity to discover novel libraries of AIEEgens that contain the minimum requirement necessary for AIEE behavior, namely, a single ring.

Meanwhile, we found that certain 1,4-DHPs could stain the mitochondria in live cells with high selectivity but without obvious guiding units (such as cationic groups). Taking one of the 1,4-DHPs as an example, we found that it exhibited excellent photostability and storage stability and that it could be utilized in applications such as real-time imaging, long-term tracking of mitochondrial morphological changes, and viscosity mapping (**Figure 55**). We believe that the use of biocatalysis

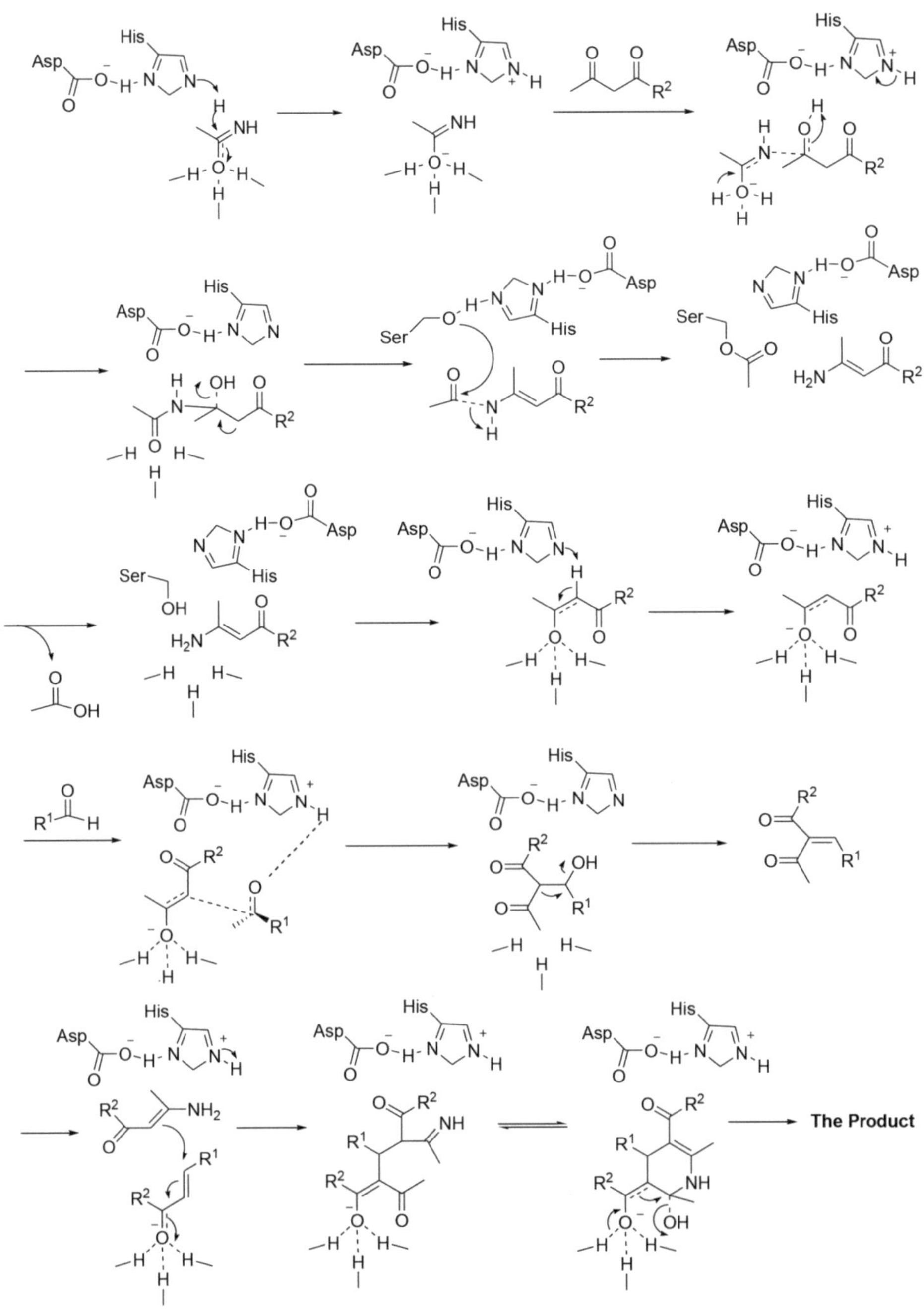

Figure 53.
Proposed mechanism of lipase-catalyzed Hantzsch-type reaction of an aldehyde with acetamide and 1,3-dicarbonyl compounds.

could be simplified with workup procedures and could provide avenues for discovering a wide variety of new materials.

In recently, Ye et al. reported solvent-free quick synthesis of 1,4-DHP calcium antagonists felodipine, nitrendipine, nifedipine, and nemadipine B and their derivatives by Lipozyme® RM IM-catalyzed multicomponent reactions of aromatic aldehyde, alkyl acetoacetate, and alkyl 3-aminocrotonate under ball milling conditions (**Figure 56**) [71]. The product was obtained in moderate yield (up to 86.8%), and the effects of the reaction conditions were investigated, including catalyst loading,

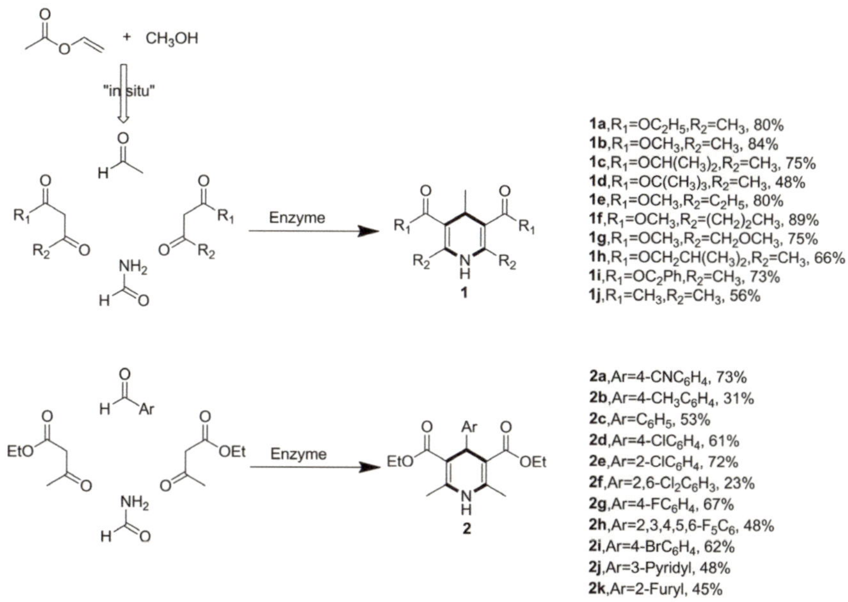

Figure 54.
Synthesis of 1 and 2 through one-pot multicomponent reactions.

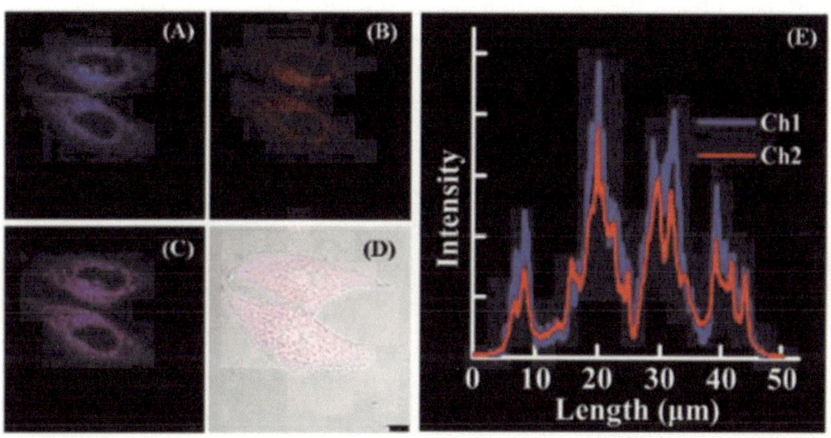

Figure 55.
Confocal fluorescence images of HeLa cells stained with 2 h and MitoTracker® Deep Red FM (MTDR). (A) Fluorescent image of 2h (10.0 μM) in HeLa cells, collected at 410-460 nm and λex = 405 nm. (B) Fluorescent image of MTDR (1.0 μM), collected at 660-740 nm and λex = 633 nm. (C) Merged image of A and B. (D) Overlay of fluorescence and bright-field image. (E) Intensity profiles of linear regions of interest (ROI) across the HeLa cells. Scale bar: 7.5 μm.

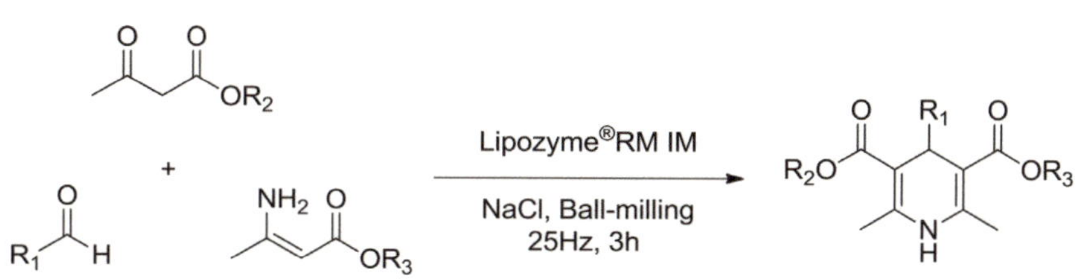

Figure 56.
Lipozyme® RM IM-catalyzed rapid synthesis of 1,4-DHP calcium antagonists and derivatives under ball milling conditions.

grinding aid, and milling frequency. The reaction features environmentally friendly, simple, and efficient operation. A major feature distinguishing this enzyme promiscuity from previously reported work is the use of mechanochemical ball milling techniques that overcome disadvantages such as long reaction times and the use of hazardous organic solvents. This work demonstrates the potential application of mixed enzyme-catalyzed reactions for drug synthesis under ball milling conditions.

4.4 Ugi reaction

In 2013, Berlozecki et al. reported for the first time an enzyme-catalyzed Ugi reaction that has many advantages over previous reactions, such as good reaction at room temperature and extensive solvent selection (**Figure 57**) [72]. In this three-component reaction, the aldehyde, amine, and isocyanide are condensed to form a dipeptide. Of all the selected lipases, Novozym 435 had the highest yield of 75%.

Recently, Thomas et al. reported the concatenation of the Ugi four-component synthesis, and the CALB-catalyzed aminolysis of the intermediary formed Ugi methyl ester products furnishes a novel consecutive five-component reaction for the formation of triamides (**Figure 58**) [73]. This one-pot method is compatible with metal catalysis methods such as copper-catalyzed alkyl azide ring addition and Suzuki cross-coupling or both in a one-pot process.

The mild reaction conditions make this sequence superior to the stepwise process with isolation of the Ugi product and even more favorable than other base-catalyzed or microwave-assisted aminolyses. This efficient scaffold forming

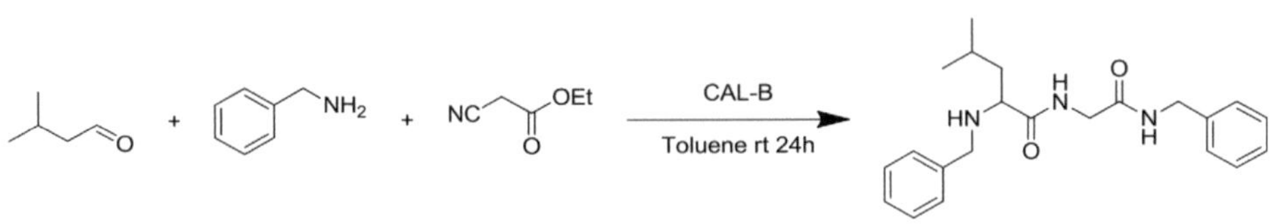

Figure 57.
CALB-catalyzed Ugi reaction.

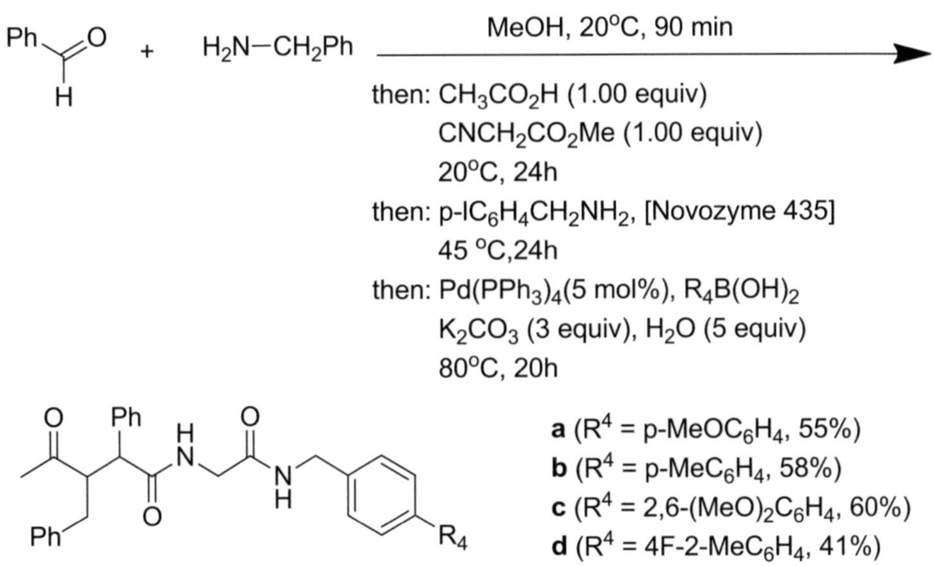

Figure 58.
Consecutive six-component U-4CR-CALB-catalyzed aminolysis-Suzuki cross-coupling sequence of biaryls.

processes are particularly favorable for creating compound libraries for medicinal chemistry lead finding and for functional chromophores in materials sciences.

5. Summary

This chapter has reviewed some examples of various types of hydrolase catalytic promiscuous reactions and their applications in the past decade. Several different types of hydrolases catalyzed carbon-carbon or carbon-heteroatom formation reactions have been discussed: aldol reactions, Michael reactions, and multicomponent reactions.

From these examples, it is clear that enzymes that display catalytic promiscuity can provide new opportunities for organic synthesis. Exploiting enzyme catalytic promiscuous reactions might lead to new, efficient, and stable catalysts with alternative activity and could provide more promising and green synthetic methods for organic chemistry. The development of protein engineering and enzyme engineering can extend the application of metagenome libraries and find enzymes with specific promiscuous behavior. We believe the progress in the area of biocatalytic promiscuity will greatly extend the useful applications of enzymes.

Acknowledgements

The authors gratefully acknowledge funding of our research in this area by grants from the Natural Science Foundation of Guangdong Province (Grant No. 2018A030307022) and the Special Innovation Projects of Common Universities in Guangdong Province (Grant No.2018KTSCX126).

Author details

Yun Wang[1] and Na Wang[2]*

1 School of Chemistry and Chemical Engineering, Lingnan Normal University, Zhanjiang, China

2 Key Laboratory of Green Chemistry and Technology, Ministry of Education, College of Chemistry, Sichuan University, Chengdu, P.R. China

*Address all correspondence to: wnchem@scu.edu.cn

References

[1] Faber K. Biotransformations in Organic Chemistry. 7th ed. Berlin Heidelberg: Springer-Verlag; 2018. 434 p. DOI: 10.1007/978-3-319-61590-5

[2] van Rantwijk F, Sheldon RA. Biocatalysis in ionic liquids. Chemical Reviews. 2007;**107**(6):2757-2785. DOI: 10.1021/cr050946x

[3] Yadav JS, Reddy GSKK, Sabitha G, Krishna AD, Prasad AR, Hafeez URR, et al. Daucus carota and baker's yeast mediated bio-reduction of prochiral ketones. Tetrahedron: Asymmetry. 2007;**18**(6):717-723. DOI: 10.1016/j.tetasy.2007.03.009

[4] Schmid A, Dordick JS, Hauer B, Kiener A, Wubbolts M, Witholt B. Industrial biocatalysis today and tomorrow. Nature. 2001;**409**:258-268. DOI: 10.1038/35051736

[5] Ran N, Zhao L, Chen Z, Tao J. Recent applications of biocatalysis in developing green chemistry for chemical synthesis at the industrial scale. Green Chemistry. 2008;**10**(4): 361-372. DOI: 10.1039/b716045c

[6] Koeller KM, Wong CH. Enzymes for chemical synthesis. Nature. 2001;**409**: 232-240. DOI: 10.1038/35051706

[7] Liese A, Seelbach K, Wandrey C, editors. Industrial Biotransformations, 2nd, Completely Revised and Extended Edition. Weinheim: Wiley-VCH; 2006. 536 p. DOI: 10.1002/3527608184

[8] O'Brien PJ, Herschlag D. Catalytic promiscuity and the evolution of new enzymatic activities. Chemistry & Biology. 1999;**6**:R91-R105. DOI: 10.1016/S1074-5521(99)80033-7

[9] Copley S. Enzymes with extra talents: Moonlighting functions and catalytic promiscuity. Current Opinion in Chemical Biology. 2003;**7**(2):265-272. DOI: 10.1016/S1367-5931(03)00032-2

[10] Bornscheuer UT, Kazlauskas RJ. Catalytic promiscuity in biocatalysis: Using old enzymes to form new bonds and follow new pathways. Angewandte Chemie International Edition. 2004; **43**(45):6032-6040. DOI: 10.1002/anie.200460416

[11] Kazlauskas RJ. Enhancing catalytic promiscuity for biocatalysis. Current Opinion in Chemical Biology. 2005; **9**(2):195-201. DOI: 10.1016/j.cbpa.2005.02.008

[12] Hult K, Berglund P. Enzyme promiscuity: Mechanism and applications. Trends in Biotechnology. 2007;**25**(5):231-238. DOI: 10.1016/j.tibtech.2007.03.002

[13] Neuberg C, Hirsch J. An enzyme which brings about union into carbon chains. Biochemische Zeitschrift. 1921; **115**:282-310

[14] Miao YF, Rahimi M, Geertsema EM, Poelarends GJ. Recent developments in enzyme promiscuity for carbon–carbon bond-forming reactions. Current Opinion in Chemical. 2015;**25**:115-123. DOI: 10.1016/j.cbpa.2014.12.020

[15] López-Iglesias M, Gotor-Fernández V. Recent advances in biocatalytic promiscuity: Hydrolase-catalyzed reactions for nonconventional transformations. The Chemical Record. 2015;**15**:743-759. DOI: 10.1002/tcr.201500008

[16] Kitazume T, Ikeya T, Murata K. Synthesis of optically active trifluorinated compounds: Asymmetric Michael addition with hydrolytic enzymes. Chemical Communications. 1986;**17**:1331-1333. DOI: 10.1039/c39860001331

[17] Cai Y, Yao SP, Wu Q, Lin XF. Michael addition of imidazole with acrylates catalyzed by alkaline protease from Bacillus subtilis in organic media. Biotechnology Letters. 2004;**26**:525-528. DOI: 10.1023/b:bile.0000019562.21256.39

[18] Torre O, Alfonso I, Gotor V. Lipase catalysed Michael addition of secondary amines to acrylonitrile. Chemical Communications. 2004;**15**:1724-1725. DOI: 10.1039/b402244k

[19] Dhake KP, Tambade PJ, Singhal RS, Bhanage BM. Promiscuous candida antarctica lipase B-catalyzed synthesis of b-amino esters via aza-Michael addition of amines to acrylates. Tetrahedron Letter. 2010;**51**:4455-4458. DOI: 10.1016/j.tetlet.2010.06.089

[20] López-Iglesias M, Busto E, Gotor-Fernández V, Gotor V. Use of protease from bacillus licheniformis as promiscuous catalyst for organic synthesis: Applications in C-C and C-N bond formation reactions. Advanced Synthesis & Catalysis. 2011;**353**: 2345-2353. DOI: 10.1002/adsc.201100347

[21] Monsalve LN, Gillanders F, Baldessari A. Promiscuous behavior of Rhizomucor miehei lipase in the synthesis of N-substituted β-amino esters. European Journal of Organic Chemistry. 2012;**6**:1164-1170. DOI: 10.1002/ejoc.201101624

[22] Bonte S, Ghinea IO, Baussanne I, Xuereb JP, Dinica R, Demeunynck M. Investigation of the lipase-catalysed reaction of aliphatic amines with ethyl propiolate as a route to N-substituted propiolamides. Tetrahedron. 2013;**69**: 5495-5500. DOI: 10.1016/j.tet.2013.04.093

[23] Steunenberg P, Sijm M, Zuilhof H, Sanders JPM, Scott EL, Franssen MCR. Lipase-catalyzed aza-Michael reaction on acrylate derivatives. The Journal of Organic Chemistry. 2013;**78**:3802-3813. DOI: 10.1021/jo400268u

[24] Rivera-Martınez JD, Escalante J, Lopez-Munguıa A, Marty A, Castillo E. Thermodynamically controlled chemoselectivity in lipase-catalyzed aza-Michael additions. Journal of Molecular Catalysis B: Enzymatic. 2015; **112**:76-82. DOI: 10.1016/j.molcatb.2014.12.009

[25] Ortega-Rojas MA, Rivera-Ramírez JD, Ávila-Ortiz CG, Juaristi E, González-Muñoz F, Castillo E, et al. One-pot lipase-catalyzed enantioselective synthesis of (R)-(−)-N-benzyl-3-(benzylamino)butanamide: The effect of solvent polarity on enantioselectivity. Molecules. 2017;**22**:2189-2196. DOI: 10.3390/molecules22122189

[26] Zhou LH, Wang N, Zhang W, Xie ZB, Yu XQ. Catalytical promiscuity of a-amylase: Synthesis of 3-substituted 2H-chromene derivatives via biocatalytic domino oxa-Michael/aldol condensations. Journal of Molecular Catalysis B: Enzymatic. 2013;**91**:37-43. DOI: 10.1016/ j.molcatb.2013.02.001

[27] Gu B, Hu ZE, Yang ZJ, Li J, Zhou ZW, Wang N, et al. Probing the mechanism of CAL-B-catalyzed aza-Michael addition of aniline compounds with acrylates using mutation and molecular docking simulations. ChemistrySelect. 2019;**4**:3848-3854. DOI: 10.1002/slct.201900112

[28] Madalinska L, Kwiatkowska M, Cierpał T, Kiełbasinski P. Investigations on enzyme catalytic promiscuity: The first attempts at a hydrolytic enzyme-promoted conjugate addition of nucleophiles to a,b-unsaturated sulfinyl acceptors. Journal of Molecular Catalysis B: Enzymatic. 2012;**81**:25-30. DOI: 10.1016/j.molcatb.2012.05.002

[29] Rizzo PVS, Boarin LA, Freitas IOM, Gomes RS, Beatriz A, Rinaldi AW, et al. The study of biocatalyzed thio-Michael

reaction: A greener and multi-gram protocol. Tetrahedron Letter. 2014;**55**: 430-434. DOI: 10.1016/j.tetlet.2013.11.047

[30] Cai JF, Guan Z, He YH. The lipase-catalyzed asymmetric C–C Michael addition. Journal of Molecular Catalysis B: Enzymatic. 2011;**68**:240-244. DOI: 10.1016/j.molcatb.2010.11.011

[31] Xie BH, Guan Z, He YH. Promiscuous enzyme-catalyzed Michael addition: Synthesis of warfarin and derivatives. Journal of Chemical Technology and Biotechnology. 2012;**87**: 1709-1714. DOI: 10.1002/jctb.3830

[32] Wu MY, Li K, He T, Feng XW, Wang N, Wang XY, et al. A novel enzymatic tandem process: Utilization of biocatalytic promiscuity for high stereoselective synthesis of 5-hydroxyimino-4,5-dihydrofurans. Tetrahedron. 2011;**67**:2681-2688. DOI: 10.1016/j.tet.2011.01.060

[33] Zhou LH, Wang N, Chen GN, Yang Q, Yang SY, Zhang W, et al. Lipase-catalyzed highly diastereoselective direct vinylogous Michael addition reaction of a, a-dicyanoolefifins to nitroalkenes. Journal of Molecular Catalysis B: Enzymatic. 2014;**109**:170-177. DOI: 10.1016/j.molcatb.2014.09.001

[34] Jiang L, Wang B, Li RR, Shen S, Yu HW, Ye LD. Catalytic promiscuity of Escherichia coli BioH esterase: Application in the synthesis of 3,4-dihydropyran derivatives. Process Biochemistry. 2014;**49**:1135-1138. DOI: 10.1016/j.procbio.2014.03.020

[35] Dean SM, Greenberg WA, Wong CH. Recent advances in aldolase-catalyzed asymmetric synthesis. Advanced Synthesis and Catalysis. 2007;**349**(8–9):1308-1320. DOI: 10.1002/adsc.200700115

[36] Gijsen HJM, Qiao L, Fitz W, Wong CH. Recent advances in the chemoenzymatic synthesis of carbohydrates and carbohydrate mimetics. Chemical Reviews. 1996;**96**: 443-474. DOI: 10.1021/cr950031q

[37] Palomo C, Oiarbide MJ, Garcia M. Current progress in the asymmetric aldol addition reaction. Chemical Society Reviews. 2004;**33**:65-75. DOI: 10.1002/chin.200421285

[38] Branneby C, Carlqvist P, Magnusson A, Hult K, Brinck T, Berglund P. Carbon–carbon bonds by hydrolytic enzymes. Journal of the American Chemical Society. 2003;**125**: 874-875. DOI: 10.1021/ja028056b

[39] Li C, Feng XW, Wang N, Zhou YJ, Yu XQ. Biocatalytic promiscuity: The first lipase-catalysed asymmetric aldol reaction. Green Chemistry. 2008;**10**(6): 616-618. DOI: 10.1039/b803406k

[40] Guan Z, Fu JP, He YH. Biocatalytic promiscuity: Lipase-catalyzed asymmetric aldol reaction of heterocyclic ketones with aldehydes. Tetrahedron Letters. 2012;**53**:4959-4961. DOI: 10.1016/j.tetlet.2012.07.007

[41] He YH, Li HH, Chen YL, Xue Y, Yuan Y, Guan Z. Chymopapain-catalyzed direct asymmetric aldol reaction. Advanced Synthesis & Catalysis. 2012;**354**:712-719. DOI: 10.1002/adsc.201100555

[42] Xie ZB, Wang N, Zhou LH, Wan F, He T, Le ZG, et al. Lipase-catalyzed stereoselective cross-aldol reaction promoted by water. ChemCatChem. 2013;**5**:1935-1940. DOI: 10.1002/cctc.201200890

[43] Wang N, Zhang W, Zhou LH, Deng QF, Xie ZB, Yu XQ. One-pot lipase-catalyzed aldol reaction combination. Applied Biochemistry and Biotechnology. 2013;**171**:1559-1567. DOI: 10.1007/s12010-013-0435-4

[44] Majumder AB, Gupta MN. Lipase-catalyzed condensation reaction of

4-nitrobenzaldehyde with acetyl acetone in aqueous-organic cosolvent mixtures and in nearly anhydrous media. Synthetic Communications. 2014;**44**:818-826. DOI: 10.1080/00397911.2013.834059

[45] Li R, Perez B, Jian H, Gao R, Dong MD, Guo Z. Acyl-peptide releasing enzyme from Sulfolobus tokodaii (ST0779) as a novel promiscuous biocatalyst for aldol addition. Catalysis Communications. 2015;**66**:111-115. DOI: 10.1016/j.catcom.2015.03.030

[46] Xu F, Xu J, Hu YJ, Lin XF, Wu Q. One-pot bienzymatic cascade combining decarboxylative aldol reaction and kinetic resolution to synthesize chiral b-hydroxy ketone derivatives. RSC Advances. 2016;**6**: 76829-76837. DOI: 10.1039/c6ra12729k

[47] Li W, Liu DN, Geng X, Li ZQ, Gao RJ. Real-time regulation of catalysis by remote controlled enzyme-conjugated gold nanorod composites for aldol reaction-based applications. Catalysis Science & Technology. 2019;**9**: 2221-2230. DOI: 10.1039/c9cy00167k

[48] Milner SE, Moody TS, Maguire AR. Biocatalytic approaches to the Henry (nitroaldol) reaction. European Journal of Organic Chemistry. 2012;**16**:3059-3067. DOI: 10.1002/ejoc.201101840

[49] Tang RC, Guan Z, He YH, Zhu W. Enzyme-catalyzed Henry (nitroaldol) reaction. Journal of Molecular Catalysis B: Enzymatic. 2010;**63**:62-67. DOI: 10.1016/j.molcatb.2009.12.005

[50] Gao N, Chen YL, He YH, Guan Z. Highly efficient and large-scalable glucoamylase catalyzed Henry reactions. RSC Advances. 2013;**3**: 16850-16856. DOI: 10.1039/c3ra41287c

[51] Wang JL, Li XH, Xie Y, Liu BK, Lin XF. Hydrolase-catalyzed fast Henry reaction of nitroalkanes and aldehydes in organic media. Journal of Biotechnology. 2010;**145**:240-243. DOI: 10.1016/j.jbiotec.2009.11.022

[52] Xu F, Wang J, Liu B, Wu Q, Lin XF. Enzymatic synthesis of optical pure b-nitroalcohols by combining D-aminoacylase-catalyzed nitroaldol reaction and immobilized lipase PS-catalyzed kinetic resolution. Green Chemistry. 2011;**13**:2359-2361. DOI: 10.1039/c1gc15417f

[53] Le ZG, Guo LT, Jiang GF, Yiang XB, Liu HQ. Henry reaction catalyzed by lipase a from *Aspergillus niger*. Green Chemistry Letters and Reviews. 2013;**6**: 277-281. DOI: 10.1080/17518253.2013.818721

[54] Busto E, Gotor-Fernández V, Gotor V. Protein-mediated nitroaldol addition in aqueous media. Catalytic promiscuity or unspecific catalysis? Organic Process Research & Development. 2011;**15**:236-240. DOI: 10.1021/op100130c

[55] Izquierdo DF, Barbosa O, Burguete MI, Lozano P, Luis SV, Fernández-Lafuente R, et al. Tuning lipase B from Candida antarctica C–C bond promiscuous activity by immobilization on polystyrene-divinylbenzene beads. RSC Advances. 2014;**4**:6219-6225. DOI: 10.1039/c3ra47069e

[56] Xie ZB, Wang N, Jiang GF, Yu XQ. Biocatalytic asymmetric aldol reaction in buffer solution. Tetrahedron Letters. 2013;**54**:945-948. DOI: 10.1016/j.tetlet.2012.12.022

[57] Birolli WG, Fonseca LP, Porto ALM. Aldol reactions by lipase from Rhizopus niveus, an example of unspecific protein catalysis. Catalysis Letters. 2017;**147**: 1977-1987. DOI: 10.1007/s10562-017-2121-6

[58] Zhang Y, Wang N, Xie ZB, Zhou LH, Yu XQ. Ionic liquid as a

recyclable and efficient medium for lipase-catalyzed asymmetric cross aldol reaction. Journal of Molecular Catalysis B: Enzymatic. 2014;**110**:100-110. DOI: 10.1016/j.molcatb.2014.10.008

[59] González-Martínez D, Gotor V, Gotor-Fernández V. Application of deep eutectic solvents in promiscuous lipase-catalysed aldol reactions. European Journal of Organic Chemistry. 2016;**8**: 1513-1519. DOI: 10.1002/ejoc.201501553

[60] Tian XM, Zhang SQ, Zheng LY. Enzyme-catalyzed Henry reaction in choline chloride-based deep eutectic solvents. Journal of Microbiology and Biotechnology. 2016;**26**(1):80-88. DOI: 10.4014/jmb.1506.06075

[61] Li K, He T, Li C, Feng XW, Wang N, Yu XQ. Lipase-catalysed direct Mannich reaction in water: Utilization of biocatalytic promiscuity for C–C bond formation in a "one-pot" synthesis. Green Chemistry. 2009;**11**(6):777-779

[62] He T, Li K, Wu MY, Feng XW, Wang N, Wang HY, et al. Utilization of biocatalytic promiscuity for direct Mannich reaction. Journal of Molecular Catalysis B: Enzymatic. 2010;**67**(3–4): 189-194. DOI: 10.1016/j.molcatb.2010.08.004

[63] Xue Y, Li LP, He YH, Guan Z. Protease-catalysed direct asymmetric Mannich reaction in organic solvent. Scientific Reports. 2012;**2**:761-764. DOI: 10.1038/srep00761

[64] Chen XY, Chen GJ, Wang JL, Wu Q, Lin XF. Lipase/acetamide-catalyzed carbon-carbon bond formations: A mechanistic view. Advanced Synthesis & Catalysis. 2013; **355**:864-868. DOI: 10.1002/adsc.201201080

[65] Guan Z, Song J, Xue Y, Yang DC, He YH. Enzyme-catalyzed asymmetric Mannich reaction using acylase from Aspergillus melleus. Journal of Molecular Catalysis B: Enzymatic. 2015; **111**:16-20. DOI: 10.1016/j.molcatb.2014.11.007

[66] Sharma UK, Sharma N, Kumar R, Sinha AK. Biocatalysts for multicomponent Biginelli reaction: Bovine serum albumin triggered waste-free synthesis of 3,4-dihydropyrimidin-2-(1H)-ones. Amino Acids. 2013;**44**: 1031-1037. DOI: 10.1007/s00726-012-1437-1

[67] Xie ZB, Wang N, Wu WX, Le ZG, Yu XQ. Trypsin-catalyzed tandem reaction: One-pot synthesis of 3,4-dihydropyrimidin-2(1H)-ones by in situ formed acetaldehyde. Journal of Biotechnology. 2014;**170**:1-5. DOI: 10.1016/j.jbiotec.2013.10.031.

[68] Zhang W, Wang N, Yang ZJ, Li YR, Yu Y, Pu XM, et al. Lipase-initiated tandem Biginelli reactions via in situ-formed acetaldehydes in one pot: Discovery of single-ring deep blue luminogens. Advanced. Synthesis & Catalysis. 2017;**359**:3397-3406. DOI: 10.1002/adsc.201700599

[69] Wang JL, Liu BK, Yin C, Wu Q, Lin XF. *Candida antarctica* lipase B-catalyzed the unprecedented three-component Hantzsch-type reaction of aldehyde with acetamide and 1,3-dicarbonyl compounds in non-aqueous solvent. Tetrahedron. 2011;**67**:2689-2692

[70] Zhang W, Wang N, Liu YH, Jiao SY, Zhang WW, Pu XM, et al. 1,4-Dihydropyridines: Discovery of minimal AIEEgens and their mitochondrial imaging applications. Journal of Materials Chemistry B. 2017;**5**:464-469. DOI: 10.1039/c6tb02135b

[71] Jiang L, Ye LD, Gu JL, Su WK, Ye WT. Mechanochemical enzymatic synthesis of 1,4-dihydropyridine calcium antagonists and derivatives. Journal of Chemical Technology Biotechnology. 2019;**94**:2555-2560. DOI: 10.1002/jctb.6051

[72] Klossowski S, Wiraszka B, Berlozecki S, Ostaszewski R. Model studies on the first enzyme-catalyzed Ugi reaction. Organic Letters. 2013;**15**: 566-569. DOI: 10.1021/ol3033829

[73] Gesse P, Müller TJJ. Consecutive five-component Ugi-4CR-CAL B-catalyzed aminolysis sequence and concatenation with transition metal catalysis in a one-pot fashion to substituted triamides. European Journal of Organic Chemistry. 2019;**11**: 2150-2157. DOI: 10.1002/ejoc.2019 00198

Section 2

Biotechnology

Chapter 4

Effect of LPMO on the Hydrolysis of Crystalline Chitin by Chitinase A and β-*N*-Acetylglucosaminidase from *Paenibacillus* sp.

Mitsuhiro Ueda, Kei Nakadoi, Kana Tsukamoto and Shunsuke Sakurai

Abstract

We performed cloning and expression of chitinase A (Pb-ChiA), β-GlcNAcase (Pb-GlcNAcase), and lytic polysaccharide monooxygenase (Pb-LPMO) genes from *Paenibacillus* sp. The analysis of the hydrolysis products indicated Pb-ChiA to be an exo-type chitinase with 10-fold activity toward β-chitin as compared with α-chitin. The sequence of Pb-GlcNAcase was found to be similar to that of β-*N*-acetylhexosaminidase from *P. barengoltzii* (99%, WP_016313754.1). Pb-LPMO was expressed in the *Brevibacillus* expression system. Pb-ChiA was found to have affinity toward crystalline chitin higher than that of Pb-LPMO. Pb-LPMO boosted the activity of Pb-ChiA toward crystalline α-chitin but not toward crystalline β-chitin. When Pb-LPMO (3 μM) was added to the reaction mixture during the hydrolysis of crystalline α-chitin by Pb-ChiA, hydrolysis products at two-fold concentration were obtained. However, the hydrolysis products decreased upon addition of more than 3 μM Pb-LPMO to the reaction mixture.

Keywords: chitin, chitinase, lytic polysaccharide monooxygenase, β-D-*N*-acetylhexosaminidase, *Paenibacillus* sp.

1. Introduction

Chitin, which can be easily isolated from crab and shrimp shell waste, is the second most abundant biopolymer on earth. At least 10 gigatons (1×10^{13} kg) of chitin are synthesized and degraded each year in the biosphere. *N*-Acetyl-chitooligosaccharides (GlcNAc) and chitooligosaccharides (GlcN), which are hydrolyzed products of chitin and chitosan, have various biological roles and many possible future applications [1–3]. GlcNAc and glucosamine are ordinarily utilized for symptom relief in osteoarthritis patients [4]. Hyaluronic acid as heterogeneous polysaccharides contained GlcNAc, and addition of GlcNAc to cultured keratinocytes facilitates the production of hyaluronic acid in a dose-dependent manner [5].

Biotechnological tools for generating enzymes are required for environmentally sustainable production of chitin oligosaccharide, GlcNAc, or both. Multiple chitinases have been characterized and are categorized into four enzyme families, namely, families 18, 19, 23, and 48 [6, 7]. The glycoside hydrolase (GH) families 18 and 19 chitinases have almost no amino acid sequence similarities and have entirely different three-dimensional structures. Ra-ChiC from *Ralstonia* sp. A-471 was the first chitinase of the GH family 23 [7] and has been categorized to goose-type lysozyme. However, although Ra-ChiC shows chitinase activity, no activity toward peptidoglycan has been described [7]. β-GlcNAcase (exo-type enzyme) belongs to GH family 20. Chitin oligosaccharides such as *N*-acetylchitobiose and *N*-acetylchitotriose are good substrates for β-GlcNAcase [8]. Lytic polysaccharide monooxygenases (LPMOs, auxiliary activity [AA] family 10) are copper-dependent and cleave polysaccharide chains embedded in the crystalline regions of the substrate that are generally inaccessible to glycoside hydrolases [9]. Hemsworth et al. reported that fungi in particular makes considerable use of LPMOs in biomass degradation, producing more LPMO enzymes than cellulases [10]. Enzyme cocktails contained cellulases and LPMO enzymes can be utilized in the bio-refinery for lignocellulose deconstruction [10].

To understand the synergic activity of LPMO during the hydrolysis of crystalline chitin by chitinase A (GH family 18) and β-GlcNAcase from *Paenibacillus* sp., cloning and expression of chitinase A (Pb-ChiA), β-GlcNAcase (Pb-GlcNAcase), and LPMO (Pb-LPMO) genes from *Paenibacillus* sp. were carried out in this study.

2. Materials and methods

2.1 Enzyme and protein measurements

The chitinase activity was assayed by determining the reducing sugars released from soluble chitin. The activity was assayed according to a previously paper [11].

The β-GlcNAcase activity was measured by determining the release of *p*-nitrophenol from *p*NP-GlcNAc. The enzyme activity required to form 1 μmol *p*-nitrophenol was regarded as one unit. The enzyme activity was measured according to a standard assay method. The reaction mixtures comprised 40 μL of 1 mM *p*NP-GlcNAc solution in 0.1 M acetate buffer (pH 6.0) and 20 μL of enzyme solution. After the reaction mixtures were incubated for 15 min at 37°C, the enzyme reaction was stopped by adding 200 μL of 0.5 M Na_2CO_3 solution.

The LPMO activity was measured by detecting the oxidized products released from chitin. Solutions of α or β-chitin (5 mg/mL), Pb-LPMO (1.0 mM), ascorbic acid (1.0 mM), and 20 mM acetate buffer (pH 5.0) were mixed to 1.0 mL volume. The reaction mixture was incubated at 37°C for 24 h. The oxidized products were then applied onto a TSK-gel Amide-80 column (4.6 × 250 mm, Tosoh Co., Tokyo, Japan) and eluted with 70% acetonitrile at a flow rate of 0.7 mL/min; the products were monitored by absorbance measurement at 210 nm.

Protein contents were assayed using Micro BCA protein assay kits (ThermoFisher Scientific) with bovine serum albumin as the standard. Protein concentrations of Pb-LPMO, Pb-ChiA, and Pb-GlcNAcase were calculated from absorbance measurement at 280 nm and the protein extinction coefficient according to the method of Gill and von Hippel [12].

2.2 Cloning and expression of chitinase A (Pb-ChiA) gene from *Paenibacillus* sp.

Paenibacillus sp. A-471 was used as a DNA donor. *Paenibacillus* sp. A-471 is closely related to *Paenibacillus barengoltzii* G22 strain. The genome sequence of *P. barengoltzii* G22 strain has been identified [13]. To clone the chitinase, β-*N*-acetylglucosaminidase, and LPMO genes from *Paenibacillus* sp. A-471, we used the nucleotide sequences of those genes from *P. barengoltzii* G22 (EOS56884.1) as templates. (The sequence of chitinase gene from *P. barengoltzii* G22 was similar to that of chitinase A from *Bacillus circulanse* WL-12 [AAA81528.1]).

First, forward (5′-ATGCATTCGAAGAGAACCCATAGCTTCAC-3′) and reverse (5′-TTATAGCGCCTGAAATAATGCGGGCAC-3′) primers for PCR were synthesized on the regions corresponding to amino acid residues 1–10 and 689–696 of Pb-ChiA, respectively. Second, to insert the nucleotide sequence corresponding to the mature Pb-ChiA in the pColdI expression vector (Takara Bio, Kyoto, Japan), forward (5′-TATACATATGCAACCGAAAGCCGCTGAGGC; the *Nde*I site is underlined) and reverse (5′-TATATCTAGATTATAGCGCCTGAAATAATGCGGG-3′; the *Xba*I site is underlined) primers for PCR were synthesized on regions corresponding to amino acid residues 33–39 and 690–696 of Pb-ChiA, respectively. PCR was carried out in a reaction mixture (20 μL) containing the *Paenibacillus* sp. A-471 genome DNA, 0.5 μM of each primer, and 10 μL of Takara PrimeSTAR Max premix (Takara Bio), according to the protocol of Takara Bio. A DNA fragment of 2.0 kb got by PCR was cloned into the *Nde*I and *Xba*I sites of the pCold I expression vector. The nucleotides of the amplified fragment were made more firm by sequencing after ligation. This expression plasmid coded mature Pb-ChiA was named pCold-Pb-ChiA.

pCold-Pb-ChiA was transformed into *Escherichia coli* DH5α, and the transformant was cultured in an LB medium (500 mL) containing ampicillin at 37°C to an OD_{600} value of 0.4 and established by induction with IPTG (0.1 mM) at 15°C overnight. The cultured *E. coli* cells were collected by centrifugation (9000 × *g*, 20 min, 4°C) and were resuspended in 20 mM Tris–HCl buffer (pH 7.5), sonicated, and centrifuged (9000 × *g*, 20 min, 4°C) again. The supernatant was put onto a HisTrap FF column (GE Healthcare UK Ltd., Little Chalfont, Buckinghamshire, England) equilibrated with 20 mM Tris–HCl buffer (pH 7.5) containing 20 mM imidazol and 300 mM NaCl. The enzyme was eluted with a gradient of 20–300 mM imidazol in the buffer. Active fractions were collected and utilized as the purified enzyme solution.

The enzyme activity was determined using soluble chitin (40% deacetylated chitin) as a substrate under various conditions of pH and temperature. The buffer systems were as follows: pH 2.0–4.0 0.2 M glycine-HCl, pH 4.0–6.0 0.2 M acetate buffer, pH 6.0–8.0 0.2 M KH_2PO_4-K_2HPO_4, pH 8.0–9.0 0.2 M Tris–HCl, and pH 9.0–11.0 0.2 M glycine-NaOH. The effect of temperature on the enzyme activity was examined at 10–80°C.

To recognize splitting patterns from hydrolysis products of the purified recombinant enzyme, 5.0 mM *N*-acetylglucosamine oligosaccharide substrates $(GlcNAc)_{2–6}$ were dissolved in 50 mM sodium acetate buffer (pH 6.0), and aliquots of the enzyme solution were added to 400 μL of each substrate solution. Enzyme reactions were carried out at 37°C for several times, and parts of the reaction mixture were then picked up and mixed with the same volume of chilled acetonitrile (−20°C) to stop the reaction.

Analysis of hydrolysis product using HPLC carried out according to a previously paper [13].

2.3 Cloning and expression of β-GlcNAcase (Pb-β-GlcNAcase) gene from *Paenibacillus* sp.

First, forward (5′-GAGGAGGAGGTTCACGGAGGAGGAGGTTCACGG-3′) and reverse (5′-GCGAGCTGTTGGAGAAGTACTCGTA-3′) primers for PCR were respectively synthesized on the 5′-upstream and 3′-downstream regions of Pb-β-GlcNAcase. Second, to insert the nucleotide sequence of Pb-LPMO in the pCold I expression vector (Takara Bio), forward (5′-GAGCTCGGTACCCTCATGAAGCTTTTTTTT; the *Nde*I site is underlined) and reverse (5′-GATTACCTATCTAGATTACATCGACAGCGA-3′; the *Xba*I site is underlined) primers for PCR were synthesized on the region corresponding to amino acid residues 1–5 and 638–641 of Pb-β-GlcNAcase, respectively. PCR was carried out in a reaction mixture (20 μL) containing the *Paenibacillus* genome DNA with first amplified PCR fragment, 0.5 μM of each primer, and 10 μL of Takara PrimeSTAR Max premix (Takara Bio), using the same thermocycling conditions as the Section 2.2. A DNA fragment of approximately 2.0 kb obtained by PCR was cloned into the *Xho*I and *Xba*I sites of the pCold I expression vector according to the protocol of InFusion cloning (Takara Bio). The nucleotides of the amplified fragment were confirmed by sequencing after InFusion cloning. This expression-plasmid-coded Pb-β-GlcNAcase was named pCold-Pb-β-GlcNAcase. Transformation and purification of Pb-β-GlcNAcase used the same method as those of Pb-Chi. Active fractions were dialyzed with 20 mM phosphate buffer (pH 7.0) and used as the purified enzyme solution.

The enzyme activity was determined using *p*NP-GlcNAc as a substrate under various conditions of pH and temperature. The buffer systems comprised x mL of 0.2 M boric acid and 0.05 mM citrate acid, and (200–x) mL of 0.1 M Na_3PO_4•$12H_2O$ (pH 2.0–12.0). The effect of temperature on enzyme activity was examined at 10–80°C.

The activity of the purified recombinant protein was tested using soluble chitin, *p*NP-GlcNAc, *p*NP-GalNAC, and $(GlcNAc)_{2–6}$. In the case of *p*NP substrates, the released *p*-nitrophenol was detected by absorbance measurement at 405 nm. The hydrolysis products of $(GlcNAC)_{2–6}$ were assayed by HPLC and measured as described above.

HPLC analysis of the hydrolysis products of chitin oligosaccharides $(GlcNAc)_{2–6}$ was performed by the same method as described in Section 2.2. To investigate the cleavage patterns from the hydrolysis products of the purified enzyme (130 units/mL substrate: pNP-GlcNAc), 5.0 mM $(GlcNAc)_{2–6}$ was dissolved in 50 mM sodium acetate buffer (pH 6.0), and aliquots of the enzyme solution were added to 400 μL of each substrate solution.

2.4 Cloning and expression of lytic polysaccharide monooxygenase (Pb-LPMO) gene from *Paenibacillus* sp.

pBIC3 forward (5′-AGTTCCGCATTCGCTCACGGCTGGGTGGACGG-3′) and pBIC reverse (5′-CATCCTGTTAAGCTTTTATTTCAGCAGCCACAA-3′) primers for PCR were synthesized on the regions corresponding to amino acid residues 38–43 and 445–449 of Pb-LPMO, respectively. PCR was carried out in a reaction mixture (20 μL) containing the *Paenibacillus* genome DNA, 0.5 μM of each primer, and 10 μL of Takara PrimeSTAR Max premix (Takara Bio), using the same thermocycling conditions as the Section 2.2. A DNA fragment of approximately 1.4 kb obtained by PCR was cloned into the pBIC3 expression vector. The nucleotides of the amplified fragment were confirmed by sequencing after ligation. This expression-plasmid-coded Pb-LPMO was named pBIC-Pb-LPMO.

pBIC-Pb-LPMO was transformed to *Brevibacillus choshinensis* according to the protocol of the *Brevibacillus* expression system (Takara Bio). The transformant was cultured using TMNm medium (10 g/L glucose, 10 g/L Bacto Soytone, 5 g/L tuna extract, 2.0 g/L Bacto Yeast Extract, 10 mg/L $FeSO_4$•$7H_2O$, 10 mg/L $MnSO_4$•$4H_2O$, and 1.0 mg/L $ZeSO_4$•$7H_2O$) with neomycin (final concentration of 50 μg/mL) at 30°C for 2 days. After cultivation, the culture medium was centrifuged at 20,000 × *g* for 15 min at 4°C. The supernatant was added to 80% saturated $(NH_4)_2SO_4$. A pellet was recovered by centrifugation at 20,000 × *g* for 15 min at 4°C.

The pellet was then redissolved in 20 mM Tris-HCl buffer (pH 7.5) and dialyzed against 20 mM Tris-HCl buffer (pH 7.5) for 1 day. The supernatant was applied to a chitin resin column (1 × 4 cm, Biorad Laboratory, America) equilibrated with 20 mM Tris-HCl buffer (pH 7.5). The enzyme was eluted with 20 mM acetic acid. Active fractions were dialyzed with 20 mM acetate buffer (pH 5.0) and used as the purified enzyme solution.

2.5 Optimum ratio of Pb-LPMO and Pb-ChiA during the hydrolysis of crystalline chitin

Pb-LPMO (0.01, 0.1, 0.5, 1.0, 3.0, 5.0, 10, and 30 μM), 1 μM Pb-ChiA, and 1.0 mM ascorbic acid in 20 mM acetate buffer (pH 5.0) were incubated with α-chitin at 37°C for 24 h. After the reaction, the reaction mixture was boiled for 10 min to terminate the enzyme reaction. The samples were mixed with the same volume of acetonitrile. The resultant solution was applied onto a column of TSK-GEL Amide 80 (4.6 × 250 mm, Tosoh Co., Tokyo, Japan) and eluted with 70% acetonitrile at a flow rate of 0.7 mL/min, and the products were monitored by absorbance measurement at 210 nm. In this experiment, we estimated the concentration of GlcNAc, $(GlcNAc)_2$, and total sugars (GlcNAc). The concentration of $(GlcNAc)_2$ in total sugars was calculated as twice the concentration of GlcNAc.

2.6 Optimum ratio of Pb-LPMO, Pb-ChiA, and Pb-GlcNAcase during the hydrolysis of crystalline chitin

We incubated 3.0 μM Pb-LPMO, 1 μM Pb-ChiA, Pb-GlcNAcase (0, 0.1, 1.0, 3.0, 5.0, and 10 μM), and 1.0 mM ascorbic acid in 20 mM acetate buffer (pH 5.0) with α-chitin at 37°C for 24 h. The reaction mixture was then boiled for 10 min to terminate the enzyme reaction. The samples were mixed with the same volume of acetonitrile. The resultant solution was applied onto the same column and conditions as mentioned in Section 2.4. In this experiment, we estimated the concentration of GlcNAc, $(GlcNAc)_2$, and total sugars (GlcNAc). The concentration of $(GlcNAc)_2$ in total sugars was calculated as the concentration of two GlcNAc.

2.7 Additional effect of Pb-LPMO during the hydrolysis of crystalline chitin by Pb-ChiA

We incubated 1 μM Pb-ChiA and 1.0 mM ascorbic acid with or without 3 μM Pb-LPMO in 20 mM acetate buffer (pH 5.0) with α-chitin at 37°C for 24 h. The reaction mixture was then boiled for 10 min to terminate the enzyme reaction. The samples were mixed with the same volume of acetonitrile. The resultant solution was applied onto the same column and conditions mentioned in Section 2.4. In this experiment, we investigated the total concentration of GlcNAc, $(GlcNAc)_2$, and total sugars (GlcNAc). The concentration of $(GlcNAc)_2$ in total sugars was calculated as twice the concentration of GlcNAc.

3. Results and discussion

3.1 Cloning and sequencing of Pb-ChiA

The length of the Pb-Chi gene was determined to be 2091 bp. It encodes a protein of 697 amino acids. The mRNA sequence was deposited in the GenBank database (LC573547). The amino acid sequence of Pb-Chi was similar to that of the chitinases from *P. barengoltzii* (99%, AIT70967), *Paenibacillus* sp. FPU-7 (93%, BAM67142.1), and *Brevibacillus* sp. AG162 (71%, WP_142063498.1). These chitinases belong to the GH family 18, and Pb-Chi may belong to this enzyme family. In addition, all catalytically important residues of chitinases from the GH family 18 [13] were conserved in Pb-Chi (Asp199, Asp201, and Glu203), as shown in *P. barengoltzii, Paenibacillus* sp. FPU-7, and *Brevibacillus* sp. AG162. Moreover, Pb-Chi contains a signal peptide, a catalytic domain, fibronectin type III domains, and a chitin-binding domain, as reported for chitinase A from *Bacillus circulans* WL-12 and the chitinase from *Paenibacillus* sp. FPU-7 [14, 15].

The amino acid sequence of the chitin-binding domain from Pb-Chi was similar to that of the carbohydrate-binding module (CBM) family 12. Chitinase II from *Aeromonas* sp. no. 10S-24, GH family 23 chitinase from *Ralstonia* sp. A-471, and chitinase A from *Bacillus circulans* WL-12 also belong to CBM 12 [16–18]. The aromatic amino acids (tryptophan, phenylalanine, and tyrosine) are present in the binding site of a number of sugar-binding proteins. The interaction of the saccharide with the aromatic residue is determined by their relative position and orientation, which have been found to vary in different sugar-binding proteins [19]. It is reported that CBM family 12 containing GH family 23 chitinase from *Ralstonia* sp. A-471 interacts with the insoluble chitin, but not with soluble chitin [17]. This binding domain can improve the hydrolysis ability against insoluble chitin. Watanabe et al. reported that W687 and Q679 are important amino acid residues during the interaction of saccharides with CBD from *B. circulans* WL-12 [20].

3.2 Expression of Pb-ChiA gene and characterization of recombinant protein

The molecular mass of the protein rPb-ChiA was 70 kDa, as determined by SDS-PAGE (data not shown), in good agreement with that predicted from the amino acid sequence. The activity of rPb-ChiA at various temperatures and pH values were determined in enzyme assays using soluble chitin as the substrate. The optimum pH and temperature of rPb-Chi were 6.0 (**Figure 1A**) and 50°C (**Figure 1B**), respectively.

To recognized the action modes of purified rPb-Chi, enzyme assays were carried with chitin oligosaccharides of several lengths, such as *N*-acetylchitobiose $(GlcNAc)_2$, *N*-acetylchitotriose $(GlcNAc)_3$, *N*-acetylchitotetraose $(GlcNAc)_4$, *N*-acetylchitopentaose $(GlcNAc)_5$, and *N*-acetylchitohexaose $(GlcNAc)_6$; buffer

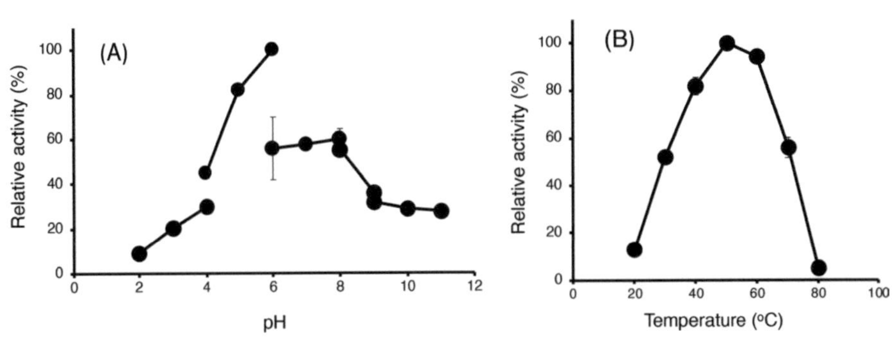

Figure 1.
Functional properties of purified Pb-ChiA. All reactions were conducted with purified enzyme and soluble chitin as the substrate. (A) Effect of pH on enzyme activity at 37°C in 0.2 M of the following buffers: Glycine-HCl (pH 2.0–4.0), sodium acetate (pH 4.0–6.0), KH_2PO_4-K_2HPO_4 *(pH 6.0–8.0), Tris-HCl (pH 8.0–9.0), and glycine-HCl (pH 9.0–11.0). (B) Effect of temperature on enzyme activity measured at 20–80°C. The average values of triplicate measurements were used as each activity values.*

aliquots were picked up over time and analyzed by HPLC (**Figure 2**). The main hydrolysis products from $(GlcNAc)_6$ were $(GlcNAc)_2$ (**Figure 2A**), and those from $(GlcNAc)_5$ were $(GlcNAc)_3$ and $(GlcNAc)_2$ (**Figure 2B**). It was suggested that $(GlcNAc)_3$ is further degraded to $(GlcNAc)_2$ and GlcNAc. The major hydrolysis products from $(GlcNAc)_4$ was $(GlcNAc)_2$ (**Figure 2C**) and those from $(GlcNAc)_3$ were $(GlcNAc)_2$ and GlcNAc (**Figure 2D**). In these experiments, $GlcNAc_2$ was not degraded, and the major hydrolysis product of colloidal chitin was $GlcNAc_2$ (data not shown). Mainly, rPb-Chi hydrolyzed the second glycosidic bond from the nonreducing end and/or reducing end, and hydrolysis patterns and products made by rPB-Chi were resemble to those described for chitinase A from *Ralstonias* sp. A-471, chitinase A from *Serratia marcescens*, and chitinase A1 from *B. circulans* WL-12 [14, 21]. The present data suggest that this enzyme is a processive exo-type chitinase.

3.3 Cloning and sequencing of Pb-GlcNAcase

The length of the Pb-GlcNAcase gene was determined to be 1926 bp. It encodes a protein of 642 amino acids. The mRNA sequence was deposited in the GenBank database (LC573548). The amino acid sequence of Pb-GlcNAcase was similar to that of the β-*N*-acetylhexosaminidase from *P. barengoltzii* (99%, WP_016313754.1), β-*N*-GlcNAcase from *Fontibacillus panacisegetis* (63%, WP_091231148.1), and β-*N*-acetylhexosaminidase from *Gorillbacterium massiliense* (49%, WP_040950886.1). These β-*N*-GlcNAcase and β-*N*-acetylhexosaminidases belong to the GH family 20, and so Pb-GlcNAcase may belong to this enzyme family. In addition, all catalytically important residues of GlcNAcase from the GH family 20 [22] were conserved in Pb-GlcNAcase (Asp228 and Glu229).

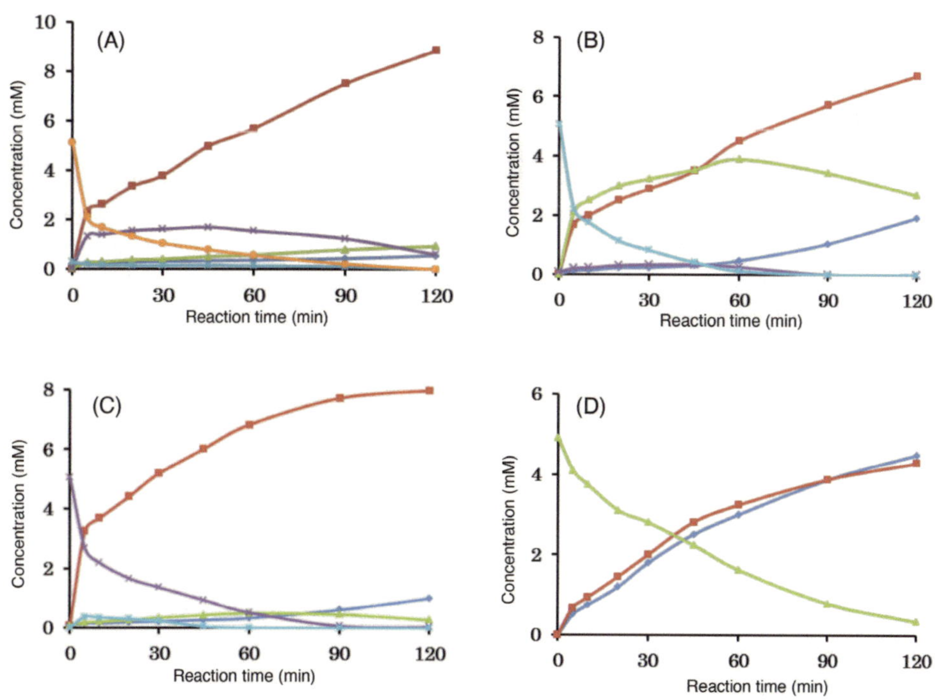

Figure 2.
HPLC analysis of products of hydrolysis of chitin oligosaccharides $(GlcNAc)_3$ to $(GlcNAc)_6$ by recombinant Pb-ChiA. The hydrolysis products of (A) $(GlcNAc)_6$, (B) $(GlcNAc)_5$, (C) $(GlcNAc)_4$, and (D) $(GlcNAc)_3$ by the enzyme were detected by HPLC, as described in the materials and methods. Lines: GlcNAc (dark-blue line), $(GlcNAc)_2$ (red line), $(GlcNAC)_3$ (yellowish olive-green line), $(GlcNAc)_4$ (purple line), $(GlcNAc)_5$ (light blue line), and $(GlcNAc)_6$ (mustard-yellow line).

3.4 Expression of Pb-GlcNAcase gene and characterization of recombinant protein

The molecular mass of the recombinant protein rPb-GlcNAcase was 70 kDa, as determined by SDS-PAGE (data not shown), in good agreement with that predicted from the amino acid sequence. The activity of rPb-GlcNAcase at various temperatures and pH values were determined in enzyme assays using *p*NP-GlcNAc as the substrate. The optimum pH of rPb-GlcNAcase was 6.0 (**Figure 3A**), while the optimal temperature for rPb-Chi activity was 50°C (**Figure 3B**); these values were similar to those of rPb-ChiA.

To recognized the action modes of purified rPb-GlcNAcase, enzyme assays were carried with chitin oligosaccharides of several lengths, such as $(GlcNAc)_2$, $(GlcNAc)_3$, $(GlcNAc)_4$, $(GlcNAc)_5$, and $(GlcNAc)_6$; buffer aliquots were picked up over time and analyzed by HPLC (**Figure 4**). The main hydrolysis products of $(GlcNAc)_6$ were GlcNAc and $(GlcNAc)_5$ (**Figure 4A**), those of $(GlcNAc)_5$ were GlcNAc and $(GlcNAc)_4$ (**Figure 4B**), and those of $(GlcNAc)_4$ and $(GlcNAc)_3$ were GlcNAc (**Figure 4C, D**). In these experiments, $(GlcNAc)_2$ was degraded to GlcNAc (**Figure 4E**). In addition, the specific activities of $(GlcNAc)_2$ and $(GlcNAc)_3$ were higher than those of $(GlcNAc)_{4–6}$ (**Table 1**). The present data indicate that this enzyme is an exo-type carbohydrate hydrolase.

3.5 Cloning and sequencing of Pb-LPMO

The length of the Pb-LPMO gene was determined to be 1350 bp. It encodes a protein of 449 amino acids. The amino acid sequence of Pb-LPMO was the same or very similar to that of the LPMO from *Paenibacillus* sp. (100%, WP_028538775) and *P. barengoltzii* (99%, WP_016312786) and the chitin-binding domain from *Paenibacillus* sp. oral taxon 786 (96%, WP_009222736). As these LPMOs belong to the AA family 10, Pb-LPMO may also belong to this enzyme family. In addition, catalytically important residues of LPMO from the AA family 10 [9] were conserved in Pb-LPMO (His38, His123) as well as the residues of LPMO from *P. barengoltzii* (WP_016312786), *Paenibacillus* sp. (WP_028538775), and *S. marcescens* BJL200 (AAU882020.1). Moreover, Pb-LPMO contains a signal peptide, an LPMO domain, two fibronectin type III domains, and a chitin-binding domain, as reported for the LPMOs of *P. barengoltzii* (WP_016312786) and *Paenibacillus* sp. (WP_028538775).

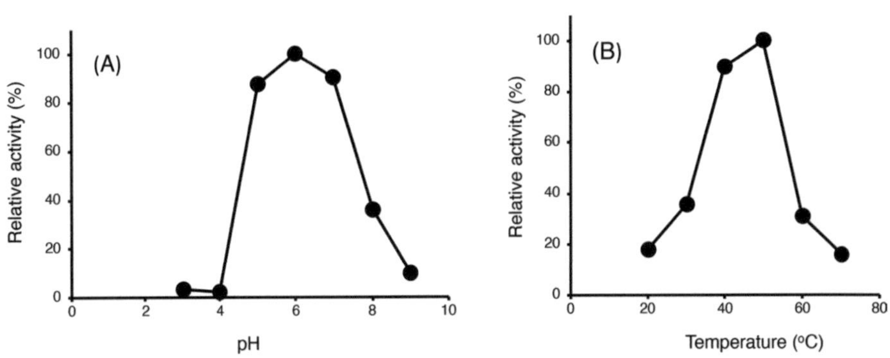

Figure 3.
Functional properties of purified Pb-GlcNAc. All reactions were conducted with purified enzyme and pNp-GlcNAc as the substrate. (A) Effect of pH on enzyme activity at 37°C in various buffers. The buffer systems comprised x mL of 0.2 M boric acid and 0.05 mM citrate acid, and (200–x) mL of 0.1 M Na_3PO_4•$12H_2O$ (pH 3.0–9.0). (B) Effect of temperature on enzyme activity measured at 20–70°C. The average values of triplicate measurements were used as activity values.

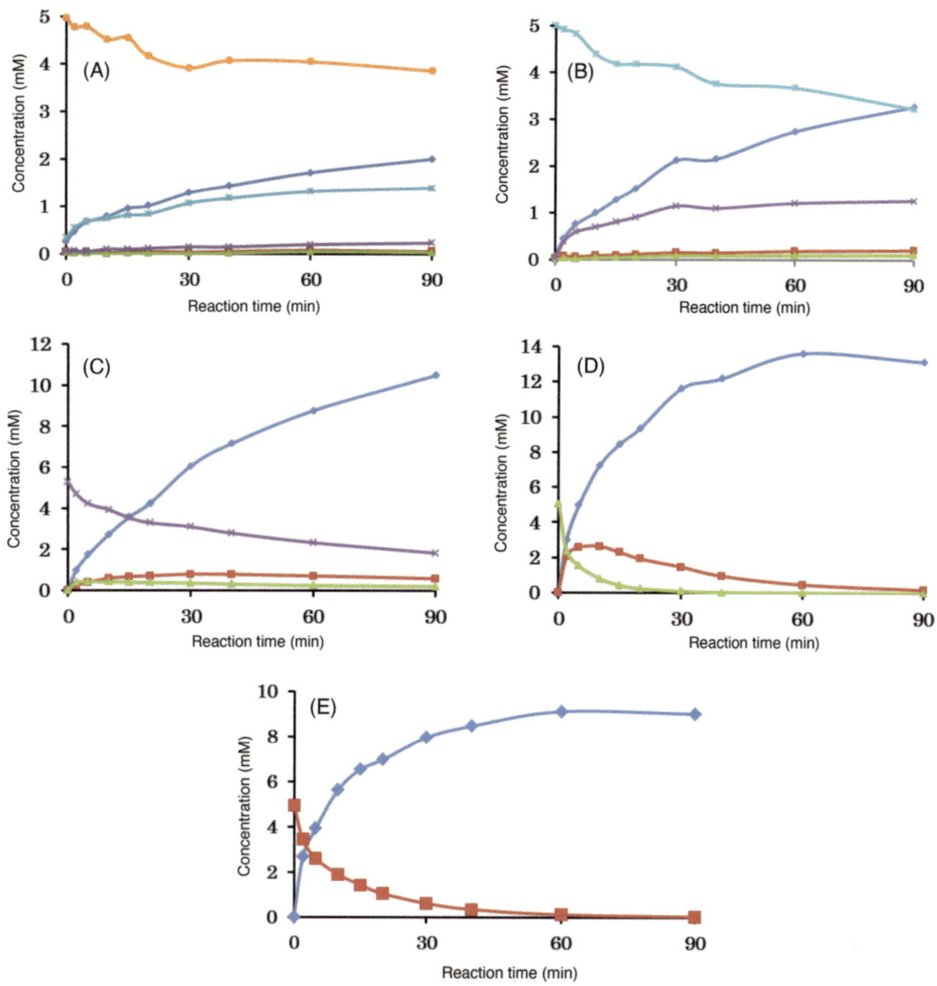

Figure 4.
HPLC analysis of products of hydrolysis of chitin oligosaccharides (GlcNAc)$_3$ to (GlcNAc)$_6$) by recombinant Pb-GlcNAcase. The degradation products of (A) (GlcNAc)$_6$, (B) (GlcNAc)$_5$, (C) (GlcNAc)$_4$, (D) (GlcNAc)$_3$, and (E) (GlcNAc)$_2$ by the enzyme were detected by HPLC, as described in the materials and methods. Lines: GlcNAc (dark-blue line), (GlcNAc)$_2$ (red line), (GlcNAC)$_3$ (yellowish olive-green line), (GlcNAc)$_4$ (purple line), (GlcNAc)$_5$ (light-blue line), (GlcNAc)$_6$ (mustard-yellow line).

Substrate	Activity (units/ mg protein)
Soluble chitin	ND
pNP-GlcNAc	111×10^{-3}
pNP-GalNAc	73.7×10^{-3}
(GlcNAc)$_2$	10.2
(GlcNAc)$_3$	12.7
(GlcNAc)$_4$	5.8
(GlcNAc)$_5$	3.8
(GlcNAc)$_6$	3.7

The average values of triplicate measurements were used as each activity value.

Table 1.
Substrate specificity of Pb-GlcNAcase.

3.6 Expression of Pb-LPMO gene and characterization of recombinant protein

The molecular mass of the recombinant protein rPb-LPMO was 45 kDa, as determined by SDS-PAGE (data not shown), in good agreement with that predicted from the amino acid sequence. The activity of rPb-LPMO was determined in enzyme assays using α-chitin or β-chitin as the substrate. After incubation, oxidized products were detected by HPLC (data not shown).

3.7 Optimum ratio of Pb-LPMO and Pb-ChiA during the hydrolysis of crystalline chitin

To understand the effect of Pb-LPMO on the hydrolysis of crystalline chitin, we added 0.01–30 μM Pb-LPMO to the reaction solution containing 1 μM Pb-ChiA. (**Figure** 5). With increasing concentration of Pb-LPMO, the concentration of the hydrolysis products (GlcNAc and $(GlcNAc)_2$) increased. When 1–10 μM of Pb-LPMO was added to the reaction mixture, products at higher concentration were obtained as compared with other concentrations. However, the addition of more than 10 μM Pb-LPMO decreased the concentration of the hydrolysis products in the reaction mixture. For efficient hydrolysis of crystalline chitin, addition of 3 μM Pb-LPMO to the reaction mixture containing 1 μM Pb-ChiA is optimal. The released products were approximately two-fold higher upon the addition of Pb-LPMO. Vaaje-Kolstad et al. reported that CBP21 (LPMO) strongly promotes hydrolysis of crystalline β-chitin by chitinase A and C from *S. marcescens* [23]. In this study, Pb-LPMO was found to efficiently promote hydrolysis of crystalline chitin.

3.8 Optimum ratio of Pb-LPMO, Pb-ChiA, and Pb-GlcNAcase during the hydrolysis of crystalline chitin

To understand the effect of Pb-GlcNAcase on the hydrolysis of crystalline chitin, we added 0.1–10 mM Pb-LPMO to the reaction solution containing

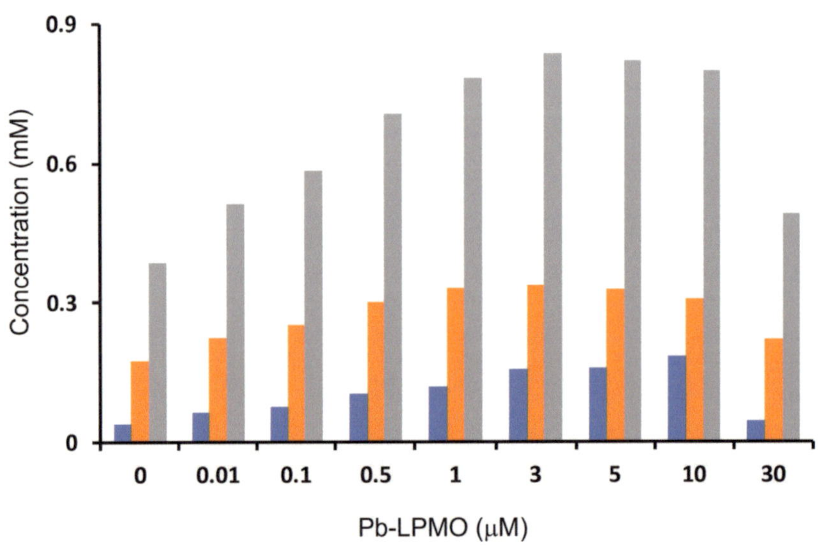

Figure 5.
Optimum ratio of Pb-LPMO and Pb-ChiA during the hydrolysis of crystalline chitin. Blue bar: GlcNAc content; reddish brown bar: $(GlcNAc)_2$ content; gray bar: Total GlcNAc content. Total GlcNAc content: $(GlcNAc)_2$ (reddish brown bar) content was converted to twice the GlcNAc content. Total GlcNAc content = GlcNAc (blue bar) content + converted GlcNAc content.

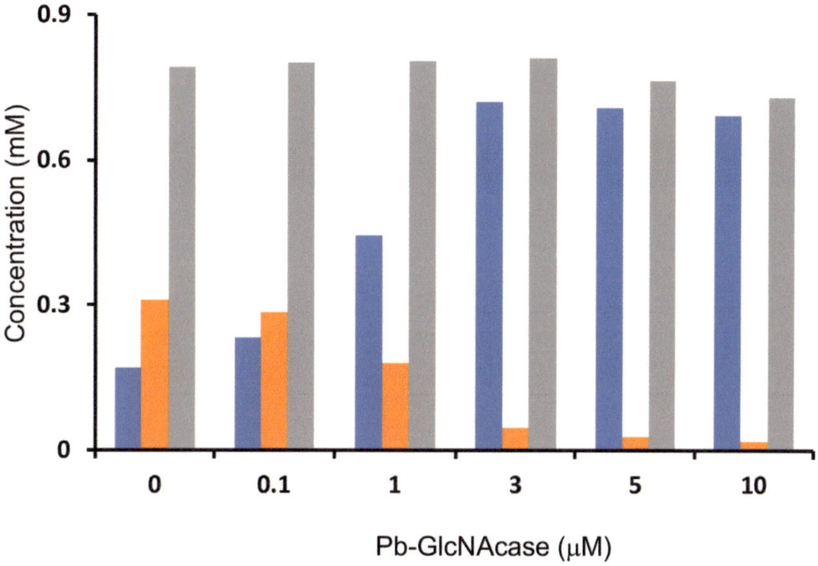

Figure 6.
Optimum ratio of Pb-LPMO, Pb-ChiA, and Pb-GlcNAcase during the hydrolysis of crystalline chitin. Blue bar: GlcNAc content; reddish brown bar: $(GlcNAc)_2$ content; gray bar: Total GlcNAc content. Total GlcNAc content: $(GlcNAc)_2$ (reddish brown bar) content was converted to twice the GlcNAc content. Total GlcNAc content = GlcNAc (blue bar) content + converted GlcNAc content.

1 mM Pb-ChiA and 3 mM Pb-LPMO (**Figure 6**). With increasing concentration of Pb-GlcNAcase, the concentration of released GlcNAc increased, but the concentration of released $(GlcNAc)_2$ decreased. It is possible that $(GlcNAc)_2$ was hydrolyzed by Pb-GlcNAcase. The concentration (approximately 0.8 mM) of released products (**Figure 6**, gray bar) when Pb-GlcNAcase was added was almost the same as that obtained from the reaction of only Pb-ChiA and Pb-LPMO (**Figure 5**). Therefore, Pb-GlcNAcase has no effect on the hydrolysis of crystal chitin but only contributes to the degradation of chitin oligosaccharides. Wang et al. reported that only β-*N*-GlcNAcase plays a role in the hydrolysis of chitin-oligosaccharides and transglycosylation [24]. Hemsworth et al. reported that it will become increasingly important to understand how cellulase and LPMO enzymes synergize with one another in biomass breakdown [10]. It found that synergization of chitinase and LPMO enzyme is also very important in the crystalline chitin hydrolysis.

4. Conclusion

Pb-ChiA had a higher affinity toward crystalline chitin than Pb-LPMO. Pb-LPMO boosts the activity of Pb-ChiA toward crystalline α-chitin but not toward crystalline β-chitin. For the efficient hydrolysis of crystalline chitin, addition of 3 μM Pb-LPMO to the reaction mixture containing 1 μM Pb-ChiA was optimal. When Pb-LPMO (3 μM) was added to the reaction mixture during the hydrolysis of crystalline α-chitin by Pb-ChiA, two-fold higher concentration hydrolysis products were obtained. It was found that Pb-GlcNAcase is involved only in the degradation of chitin oligosaccharides. Synergization of Pb-ChiA and Pb-LPMO was shown during crystalline chitin decomposition. LPMO contributes to the efficiency of chitinase used in the conversion of chitin biomass.

Author details

Mitsuhiro Ueda*, Kei Nakadoi, Kana Tsukamoto and Shunsuke Sakurai
Graduate School of Life and Environmental Sciences, Osaka Prefecture University, Sakai, Osaka, Japan

*Address all correspondence to: mueda@biochem.osakafu-u.ac.jp

References

[1] Suzuki K, Tokoro A, Okawa Y, Suzuki S, Suzuki M. Enhancing effects of *N*-acetyl-chitooligosaccharides on the active oxygen-generating and microbicidal activities of peritoneal exudates cells in mice. Chemical & Pharmaceutical Bulletin. 1985;**36**:886-888

[2] Tokoro A, Tatewaki N, Suzuki K, Mikami T, Suzuki S, Suzuki M. Growth-inhibitory effect of hexa-*N*-acetylchitohexaose and chitohexaose against meth-a solid tumor. Chemical & Pharmaceutical Bulletin. 1988;**36**:784-790

[3] Hirano S, Nagao N. Effects of chitosan, pectic acid, lysozyme, and chitinase on the growth of several phytopathogens. Agricultural and Biological Chemistry. 1989;**53**: 3065-3066

[4] Shikhman AR, Brinson DC, Valbrach J, Lotz MK. Differential metabolic effects of glucosamine and *N*-acetylglucosamine in human articular chondrocytes. Osteoarthritis and Cartilage. 2009;**17**:1022-1028

[5] Chen JK, Shen CR, Liu CL. *N*-Acetylglucosamine: Production and applications. Marine Drugs. 2010;**8**:2493-2516

[6] Fukamizo T. Chitinolytic enzymes: Catalysis, substrate binding, and their application. Current Protein & Peptide Science. 2001;**1**:105-124

[7] Ueda M, Ohata K, Konishi T, Sutrisno A, Okada H, Nakazawa M, et al. A novel goose-type lysozyme gene with chitinolytic activity from the moderately thermophilic bacterium *Ralstonia* sp. A-471: Cloning, sequencing, and expression. Applied Microbiology and Biotechnology. 2009;**81**:1077-1085

[8] Ueda M, Fujita Y, Kawaguchi T, Arai M. Cloning, nucleotide sequence and expression of the β-*N*-acetylglucosaminidase gene from *Aeromonas* sp. No. 10S-24. Journal of Bioscience and Bioengineering. 2002;**89**:164-169. DOI: 10.1016/S1389-1723(00)88731-9

[9] Vaaje-Kolstad G, Forsberg Z, Loose JS, Bissaro B, Eijsink VG. Structural diversity of lytic polysaccharide monooxygenases. Current Opinion in Structural Biology. 2017;**44**:67-76. DOI: 10.1016/j.sbi.2016.12.012

[10] Hemsworth GR, Johnston EM, Davies GJ, Walton PH. Lytic polysaccharide monooxygenases in biomass conversion. Trends in Biotechnology. 2015;**33**:747-761. DOI: 10.1016/j.tibtech.2015.09.006

[11] Ueda M, Shioyama T, Nakadoi K, Nakazawa M, Sakamoto T, Iwamoto T, et al. Cloning and expression of a Chitinase gene from *Eisenia fetida*. International Journal of Biological Macromolecules. 2017;**104**:1648-1655. DOI: 10.1016/j.ijbiomac.2017.03.140

[12] Gill SC, von Hippel PH. Calculation of protein extinction coefficients from amino acid sequence data. Analytical Biochemistry. 1989;**182**:319-326. DOI: 10.1016/0003-2697(89)90602-7

[13] *Paenibacillus barengoltzii* G22 Assembly and Gene Annotation [Internet]. Available from: http://bacteria.ensembl.org/Paenibacillus_barengoltzii_g22/Info/Annotation/ [Accessed: April 2017]

[14] Watanabe T, Oyanagi W, Suzuki K, Takaka H. Chitinase system of *Bacillus circulans* WL-12 and importance of Chitinase A1 in chitin degradation. Journal of Bacteriology. 1990;**172**:4017-4022. DOI: 10.1128/jb.172.7.4017-4022.1990

[15] Itoh T, Hibi T, Fujii Y, Sugimoto I, Fujiwara A, Suzuki F, et al. Cooperative degradation of chitin by extracellular and cell surface—expressed chitinases from *Paenibacillus* sp. strain FPU-7. Applied and Environmental Microbiology. 2013;**79**:7482-7490. DOI: 10.1128/AEM.02483-13

[16] Shiro M, Ueda M, Kawaguchi T, Arai M. Cloning of a cluster of Chitinase genes from *Aeromonas* sp. No. 10S-24. Biochimica et Biophysica Acta. 1996;**1305**:44-48. DOI: 10.1016/0167-4781(95)00213-8

[17] Arimori T, Kawamoto N, Shinya S, Okazaki N, Nakazawa M, Miyatake K, et al. Crystal structures of the catalytic domain of a novel glycohydrolase family 23 chitinase from *Ralstonia* Sp. A-471 reveals a unique arrangement of the catalytic residues for inverting chitin hydrolysis. The Journal of Biological Chemistry. 2013;**288**:18696-18706

[18] Hashimoto M, Ikegami T, Seino S, Ohuchi N, Fukada H, Sugiyama J, et al. Expression and characterization of the chitin-binding domain of chitinase A1 from *Bacillus circulans* WL-12. Journal of Bacteriology. 2000;**182**:3045-3054. DOI: 10.1128/jb.182.11.3045-3054.2000

[19] Sujatha MS, Sasidhar YU, Balaji PV. Energetics of galactose- and glucose-aromatic amino acid interactions: Implications for binding in galactose-specific proteins. Protein Science. 2004;**13**:2502-2514. DOI: 10.1110/ps.04812804

[20] Hara M, Sugimoto H, Uemura M, Akagi K, Suzuki K, Ikegami T, et al. Involvement of Gln679, in addition to Trp687, in chitin-binding activity of the chitin-binding domain of chitinase A1 from *Bacillus circulans* WL-12. Journal of Biochemistry. 2013;**154**:185-193. DOI: 10.1093/jb/mvt043

[21] Sutrisno A, Ueda M, Abe Y, Nakazawa M, Miyatake K. A chitinase with high activity toward partially N-acetylated chitosan from a new, moderately thermophilic, chitin-degrading bacterium, *Ralstonia* Sp. A-471. Applied Microbiology and Biotechnology. 2004;**63**:398-406. DOI: 10.1007/s00253-003-1351-2

[22] Ueda M, Fujita Y, Kawaguchi T, Arai M. Cloning, nucleotide sequence and expression of the β-*N*-acetyl-glucosaminidase gene from *Aeromonas* sp. No. 10S-24. Journal of Bioscience and Bioengineering. 2000;**89**:164-169. DOI: 10.1016/S1389-1723(00)88731-9

[23] Vaaje-Kolstad HSV, van Aalten DMF, Synstad B, Eijsink VGH. The non-catalytic chitin-binding protein CBP21 from *Serratia marcescens* is essential for chitin degradation. The Journal of Biological Chemistry. 2005;**280**:28492-28497

[24] Wang LX. The amazing transglycosylation activity of Endo-β-N-acetylglucosaminidases. Trends in Glycoscience and Glycotechnology. 2011;**23**:33-52. DOI: 10.4052/tigg.23.33

Chapter 5

Current and Prospective of Breast Cancer Biomarkers

Stephen Rathinaraj Benjamin and Fabio de Lima

Abstract

Biomarkers have shown great promise over the past decade the process of drug development more effective and have become an integral part of diagnosis of diseases. Biosensors were integrated with biomarker detection and point-of-care detection for signal amplification, high specificity and sensitivity, rapid response time, low cost, simplicity and multi-analytical testing. In order to detect more sensitively, these particular biomarkers have been explored with the possibility of real-time measurements in order to develop simple and compact systems which can analyze complex specimens. Various biosensors including electrochemical biosensors have recently been developed based on disease-specific biomarkers in the diagnosis of cancer disease. The main objective of the book chapter is to review research with new materials/methods in electrochemical biosensing techniques to detection of breast cancer biomarkers and evaluating latest techniques for detection of important analytes in real samples. In this book chapter, the recent development of electrochemical biosensors of breast cancer biomarkers will be reviewed. Furthermore, recent and future trend application of breast cancer biomarkers will be discussed.

Keywords: biomarkers, breast cancer, electrochemical immunosensors

1. Introduction

Cancer is considered a complex disease because several factors are involved the disease outbreak [1]. Breast cancer (BCa) is among the most common cancers in women, accounting for about 14% of deaths from cancer in women around the world. In both the developed and less developed world, breast cancer is the most common cancer in women. About 508,000 women worldwide are estimated to have died from breast cancer in 2011 [2].

One of the most important methods is the detection and analysis of cancer biomarkers. The development of cancer has the maximum potential for therapeutic intervention. Recent developments in molecular biology have made it clear that biomarkers of cancer play a major role in the treatment, prognosis and insight into cancer etiology. The National Cancer Institute describes a biomarker as a "biological molecule contained in blood, other body fluids or tissues that is a sign of a normal or abnormal process or disease or disorder."

Cancer markers are one of the most valuable tools for early detection, diagnosis, treatment, progression tracking and evaluation of chemotherapy resistance. There are over 200 different cancer-related diseases impacting various parts of the human body. Tumor markers are usually found at low levels in the absence of a tumor. After tumor formation, level changes and therefore cancer marker clinical tests must be fast, selective and sensitive sufficiently distinguish slight changes in marker levels in complex biological fluids.

Eventually, the extensive use of tumor markers in healthcare would rely on the identification of many highly selective and sensitive tumor markers. However, traditional immunoassays such as enzyme-linked immunosorbent assay (ELISA) have some drawbacks including time-consuming, extensive incubation procedures are performed for the antibody-antigen interaction to reach equilibrium and performed by highly skilled personnel using costly and sophisticated instruments.

1.1 Electrochemical immunosensors: definitions and methods

Over the past decade, a growing number of researchers have focused on developing fast and simple-to-use biosensor technology-based techniques to detect specific biomarkers. Under the proposed International Union of Pure and Applied Chemistry (IUPAC) concept, a biosensor combines two important components of the bioreceptor and the transducer component for target detection. The chemical biosensors have enticing instruments that can be measured to convert biological interactions into electric signals. Nevertheless, the conversion of biological signal to a measurable signal is difficult due to the limitations of the biological environment and the extremely low tumor marker rates in biological samples.

The development of effective biosensors with sensitivity and selectivity has thus gained significant emphasis not only in clinical medicine, as well as in basic medicine and biomedical engineering. It is not surprising that due to their high potential in the bioassay region, the emphasis has been shifted to electrochemical biosensors as shown in **Figure 1**.

In addition to, they promote the reuse of molecules in biorecognition and prevent a time lapse among sample preparation and analysis. Moreover, biosensors have high potential to detect multiple biomarkers simultaneously. The use of electrochemical biosensors is particularly noteworthy due to their low cost, usability, sensitivity, specificity and suitability to detect low levels of molecular biomarkers

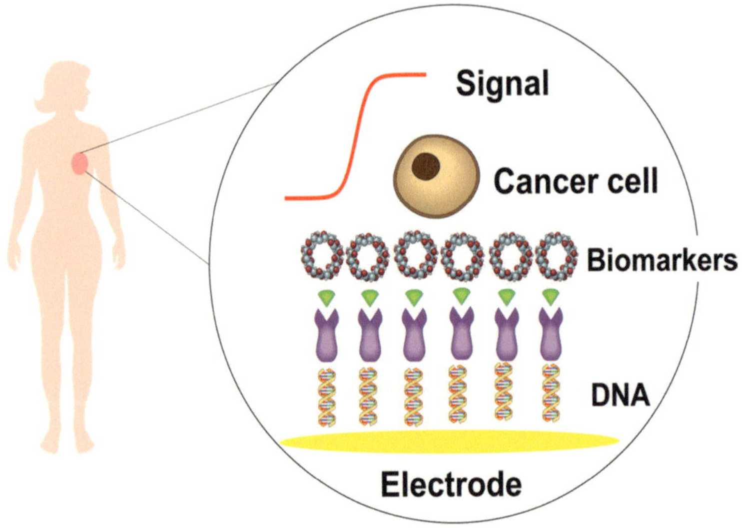

Figure 1.
The construction of breast cancer biomarkers-based biosensors.

using a range of available techniques like cyclic voltammetry (CV); differential pulse voltammetry (DPV); square-wave voltammetry (SWV) and electrochemical impedance spectroscopy (EIS). On the other hand, the electrochemical biosensors play a major role in moving towards simpler point-of-care (POC) research.

In this chapter, we addressed the molecular changes throughout cancer that have been developed and related biomarkers. Currently designed biosensor platforms and manufacturing methods to identify these biomarkers of cancer are discussed. Compared to the previous studies, this analysis highlights these biosensors analytical performance in terms of sensitivity, linear detection range and detection limit obtained with various manufacturing techniques.

2. Immunoassays with biomarkers for breast cancer

Cancer biomarkers are molecules in the presence of cancer in the body that are overexpressed. It can apply to a secreted enzyme triggered by a tumor or a specific body response to cancer. It is important to use a wide range of genomic, epigenetic, glycomic, proteomic and imaging biomarkers to recognize the point, prognosis, and epidemiology of cancer. Despite the fact that there are still multiple obstacles in turning the analysis of biomarkers into a therapeutic platform; several biomarkers dependent on genes and proteins have now been used. These biomarkers are reviewed in the following sections.

2.1 Carcinoembryonic antigen (CEA)

One of the first tumor antigens to be reported was carcinoembryonic antigen (CEA), defined in 1965. CEA is a glycoprotein of the immunoglobulin family found by radioimmunoassay or enzyme-linked immunosorbent assay in the serum of patients with cancer. A strong false positive rate in common communities and a poor test sensitivity and accuracy reduce the therapeutic benefit of CEA identification. The elevated CEA level is not unique to breast cancer because CEA can be present in many various neoplasia forms. CEA is more prevalent in ductal than in lobular carcinomas in breast tumors. During the testing process, the FDA also identified CEA as an acceptable serum biomarker for colon cancer [3]. For detection of CEA, numerous electrochemical immunosensors have been developed [4–7].

Rizwan et al. [4] fabricated the nanocomposite of gold nanoparticles (AuNPs), carbon nano-onions (CNOs), single-walled carbon nanotubes (SWCNTs) and chitosan (CS) (AuNPs/CNOs/SWCNTs/CS) for the modification of GCE (glass carbon electrode) and development of highly sensitive label-free electrochemical immunosensor for the detection of carcinoembryonic antigen (CEA), clinical tumor marker using $[Fe(CN)_6]^{3/4-}$ as mediator solution. By using layer by layer fabrication of the immunosensors were observed using CV and SWV methods. When CEA antibody combines with CEA antigen, the formed immunocomplexes formed. The decrease in the electrical signals of the immunosensor has a linear relationship for the quantitative detection of CEA ranging from 100 fg mL^{-1} to 400 ng mL^{-1} with a low detection limit of 100 fg mL^{-1}. Interestingly, a novel label-free electrochemical immunosensors [5] for detecting CEA based on gold nanoparticles (AuNPs) and Nile blue A (NB) hybridized electrochemically reduced graphene oxide (NB-ERGO) as shown in **Figure 2**. The NB-graphene oxide (NB-GO) composite was developed by the π-π interaction. The linear range of the proposed immunosensor was estimated at 0.001–40 ng mL^{-1} under optimal conditions using DPV technique and the detection limit was estimated at 0.00045 ng mL^{-1} for CEA. In addition, a silver nanoparticle (AgNPs) decorated with thionine/infinite coordination polymers as sensing platforms

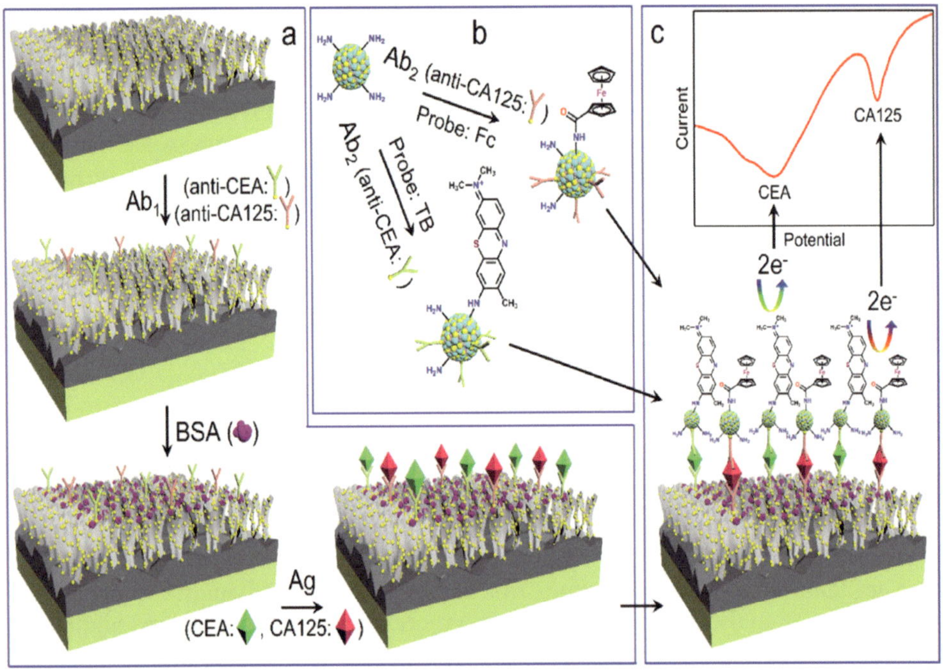

Figure 2.
Schematic of the fabrication of a sandwich-type immunosensor with a Au-VBG/BDD sensing electrode. Reprinted with permission from Ref. [5].

for detection of CEA. However, this type of glycoprotein associated with the development of breast, ovary, pancreas, lung and colon cancer. The sensor showed a detection limit of 0.5 fg mL^{-1} and a wide linear range of 50 fg mL^{-1}–100 ng mL^{-1} [6].

Additionally, simultaneous detection of carcinoembryonic antigen (CEA) and the carcinoma antigen 125 (CA125) constructed with the immunosensor containing vertical boron-doped graphene (VBG) and Boron-doped diamond (BDD) composite film by chemical vapor deposition method. These process characteristics add a wide unique surface area and strong electrocatalytic activity to the vertical BG (VBG)/BDD film resulting increase electroactive surface area. Hence, the Au-VBG/BDD signal amplification device immunosensor demonstrated strong selectivity, specificity and excellent stability for the simultaneous identification of the CEA and CA125 at 0.5–100 pg mL^{-1} and 0.5–100 mU mL^{-1} concentrations, respectively, with detection limits of 0.15 pg mL^{-1} and 0.09 mU mL^{-1} respectively [7].

2.2 Cancer antigen 15-3 (CA15-3)

CA15.3 is among the most widely accepted common biomarkers linked with breast cancer and part of the mucins family with a glycoprotein [8]. The normal level of CA15-3 is under 30 U mL^{-1} in human serum [9]. The concentration of CA15-3 is also linked to the postoperative condition, recurrence rate and monitoring of metastases of the patients. However, the serum level of this tumor marker is an important indicator for the early detection of breast cancer and the determination of disease severity. Several electrochemical biosensors focused on nanoparticles have been designed for the identification of CA15-3. According to recent studies, we can explain that the design of biosensors includes a number of distinct nanostructures among the most commonly utilized signal enhancers.

Using sandwich based electrochemical was fabricated for detection of CA15-3. According to this works, Qin and coworkers [10] used the sensitive sandwich electrochemiluminescence (ECL) immunosensors which were modified by graphene oxide-PEI carbon quantum dots (CQDs)-Au nanohybrid to detect CA15-3. Firstly, nanocomposite has been synthesized through dopamine and Ag^+ redox reaction formed in Ag nanoparticles and polydopamine (AgNPs-PDA). The high-surface nanocomposite can provide an efficient substrate for initial antibody (Ab1) immobilization. Carbon quantum dots (CQDs) are attached by amide bonds on polyethyleneimine functionalized graphene oxide (PEI-GO). Gold nanoparticles are modified on CQDs-decorated PEI-GO substrates. Then, the secondary antibody (Ab2) was immobilized by AuNPs/CQDs-PEI-GO composite. According to this report, this ECL sensor showed good linear concentration range of CA15.3 from 0.005 to 500 U mL^{-1}, with a relatively low detection limit of 0.0017 U mL^{-1}. Further, an ultrasensitive label-free electrochemical immunosensor configured with highly conductive dendritic Au@Pt core-shell nanocrystals (Au@Pt NCs) uniformly dispersed with ferrocene-grafted-chitosan (Fcg-CS) was prepared by Wang et al. [11]. Au@Pt NCs were developed using hexadecyl dimethyl benzyl ammonium chloride (HDBAC) as a growth-directing agent through a simple wet-chemical one-pot technique. According to this report, the proposed immunosensor had a low detection limit of 0.17 U mL^{-1} and worked well over a linear range of 0.5–200 U mL^{-1}. The technique designed for the immunosensor presents a feasible approach for clinical diagnostic applications.

In another study, Cobalt sulfides/graphene nanocomposite and AuNPs (CoS_2-GR-AuNPs) was used to detect CA15-3. In this composite, AuNPs act as the immobilization site for binding of antibodies. CoS_2-GR nanocomposite exhibits excellent electrocatalytic activity against catechol oxidation and also yields large surface area that enhances the amount of CA15-3 antibody immobilized as shown in **Figure 3**. The developed immunosensor showed a wide linear range of 0.1–150 U mL^{-1} and a low detection limit of 0.03 U mL^{-1}. The immunosensor demonstrated good precision, reliability, specificity and was successfully applied in serum samples for CA15-3 detection [12].

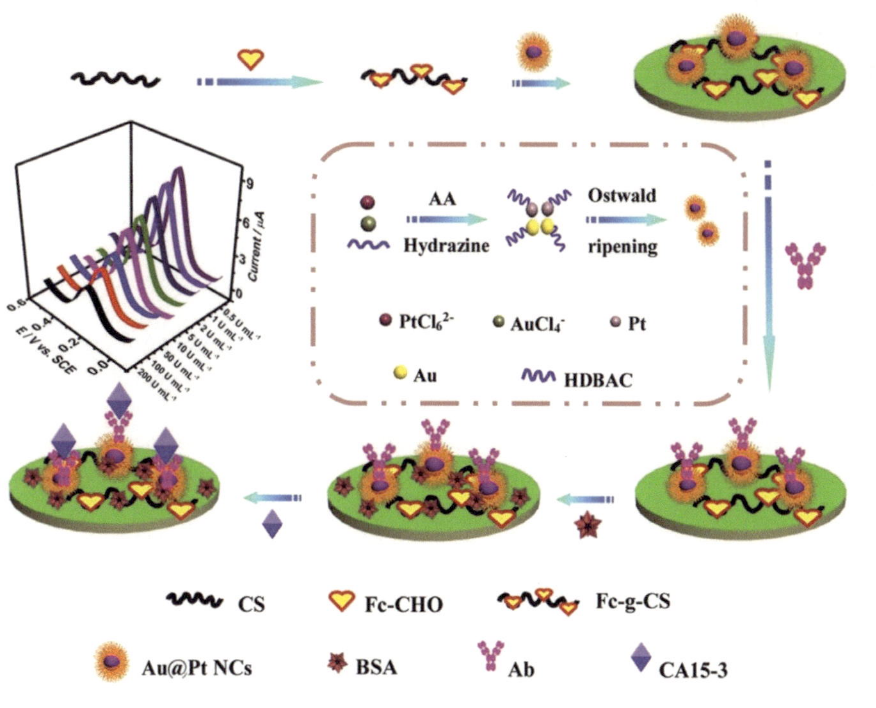

Figure 3.
Formation strategies of Fc-CS and Au@Pt NCs, along with constructing a label-free electrochemical immunosensor. Reprinted with permission from Ref. [11].

2.3 Human epidermal growth factor receptor 2 (HER2)

Anti HER2 is a monoclonal antibody (mAb) with molecular weight 185-kDa, that can bind HER2 protein and has been recognized as one of the biomarkers of breast cancer. HER2 is a receptor tyrosine kinase, a part of the cellular signaling pathways associated with the epidermal growth factor receptor (EGFR) family, which can complexes occurred in HER2 and alike proteins (such as erbB1, erbB3, and erbB4). Trastuzumab (Herceptin®), an antibody designed to treat breast cancer patients with a HER2 receptor, was the first genetically-guided medication authorized by the FDA [13], and included evaluations for its applicability before it could be used for both the diagnosis of HER2 positive gastric cancer and breast cancer [14]. HER2 overexpression in some breast cancer patients is often used as a main prognostic predictor and important treatment goals for the detection of breast cancer in adult females. The normal range in the blood of healthy women is between 2 and 15 mg L^{-1}. Since HER2 is linked with breast cancer, the development of highly sensitive biosensors to identify low levels of HER2 biomarkers is great importance.

For the detection of HER2, a disposable screen printed carbon electrode was modified with gold nanoparticles [15]. The gold nanoparticles allow rapid movement of electrons and provide a biocompatible surface to immobilize small fragments of antibodies in a directed manner, resulting in enhanced binding antigen performance. The proposed immunosensor showed a wide dynamic range of 0.01–100 ng mL^{-1} with detection limit of 0.01 ng mL^{-1}. The fabricated immunosensor with HER2-avian single chain variable fragment (ScFv) has outstanding durability with a retention rate of more than 95.6% up to 22 days. Recently, Shamsipur et al. [16] designed the label free immunosensor, functionalization of 3-aminopropyltrimethoxysilane coated magnetite nanoparticles with antibody (antiHER2/APTMS-Fe_3O_4), as a platform bioconjugate (PB), and deposited on a bare GCE. However, the PB was covered with magnetic gold nanoparticles self-assembled by thiolated antibodies (antiHER2/Hyd@AuNPs-APTMS-Fe_3O_4) containing chemically reduced silver ions, as a bioconjugate (LB) label. Under optimized conditions, using DPV, the level of HER2 was determined obtained in the range of 5.0×10^{-4}–50.0 ng mL^{-1} with a LOD of 2.0×10^{-5} ng mL^{-1}.

Furthermore, an electrochemical sandwich-based immunosensor for HER2 was designed using a lead sulfide quantum dots anti-HER2 antibody as a label (Ab2-PbS QDs) [17]. The presence of amine and hydroxyl groups from secondary anti-HER2 and coated PbS QDs are covalently linked together by carbonyldiimidazole (CDI) in the bioconjugation of PbS QDs. Using SWV signal, the proposed immunosensor, the level of HER2 was determined the linear range from 1 to 100 ng mL^{-1} with a limit of detection of 0.28 ng mL^{-1}. Recently, Freitas et al. [18] developed a quantum dots (QDs) as electrochemical label for electrochemical immunosensing strategy established on in situ detection of HER2-ECD in human serum samples. By using the screen-printed electrode, core/shell CdSe@ZnS to provide immobilization of bioreceptor functional groups as shown in **Figure 4**. The analytical performance was tested in spiked human serum samples, demonstrating an excellent performance in a wide linear range (10–150 ng mL^{-1}) with a limit of detection of 2.1 ng mL^{-1}.

2.4 MIRNAs

MIRNAs are a class of small non-coding RNA components that control the post-transcription expression of target genes by either translational repression or mRNA degradation. Mature miRNAs are made up of β22 nucleotides and are generated through 70–100 nucleotide hairpin substrate molecules. The human genome has

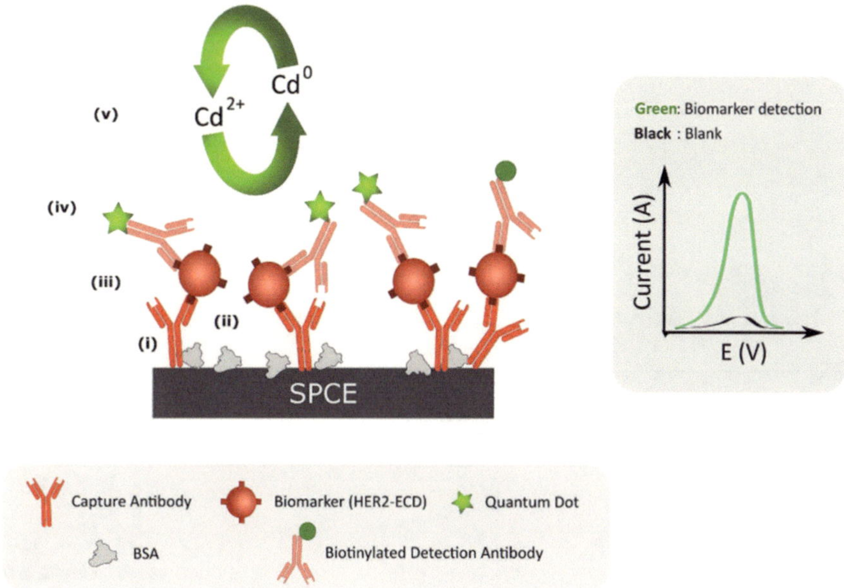

Figure 4.
Representation of the immunosensor construction and detection strategy of biomarker HER2. Reprinted with permission from Ref. [19].

been reported to be capable of encoding over 1000 miRNAs, which can control up to 60% of mammalian genes. MIRNA detection has produced a number of electrochemical sensors, which have been reviewed elsewhere.

In recent study, highly sensitive ultrasensitive electrochemical biosensor detection is of microRNA-222. In this works, the biosensor was fabricated as follows. The gold nanoparticles-modified graphene oxide (rGO/Au NPs) with capture probe (cDNA) through a thiolated group was immobilized. The modified electrode (cDNA/rGO/Au NPs/GCE) was linearly hybridized with microRNA-222 and signal probe, resulting in a sandwich structure of the modified electrode cDNA-microRNA signal probe on the surface. Under optimized conditions, using DPV technique, the biosensor of microRNA showed wide linear response range (0.5 fM to 70 nM) and a low limit of detection of 0.03 fM [19]. Further, a novel label free and simple electrochemical biosensors strategy was developed for detection of mRNA-21. From this research works, firstly confirmed with excellent ability of AuNPs superlattices for electron transport and tunable structures, a remarkable improvement was made in the immobilizing quantity of probe molecules on the electrode surface and a major improvement in the substratum's electrical signal. Moreover, toluidine blue (TB) and micro-RNA interaction established with negatively charged backbone phosphate groups increasing electrostatic interaction. By using this technique, microRNA with a relatively low detection limit of 78 aM can be identified in a linear range from 100 aM to 1 nM. The proposed electrochemical nano biosensor could be clinically valuable in early breast cancer diagnosis through direct identification of serum microRNA-21 in real clinical samples without sample testing [20].

In addition, Azimzadeh et al. [21] developed a novel electrochemical nano biosensor for the detection of miRNA-155. From this works, this methodology based on thiolated probe-functional gold nanorods (GNRs) on a graphene oxide (GO) layer on a glass carbon electrode surface as shown in **Figure 5**. Nevertheless, the authors used Oracet Blue (OB) for the first time in the proposed electrochemical nano biosensor as an electroactive miRNA label. A novel intercalating label Oracet Blue, the reduction signals were measured using the method of differential pulse

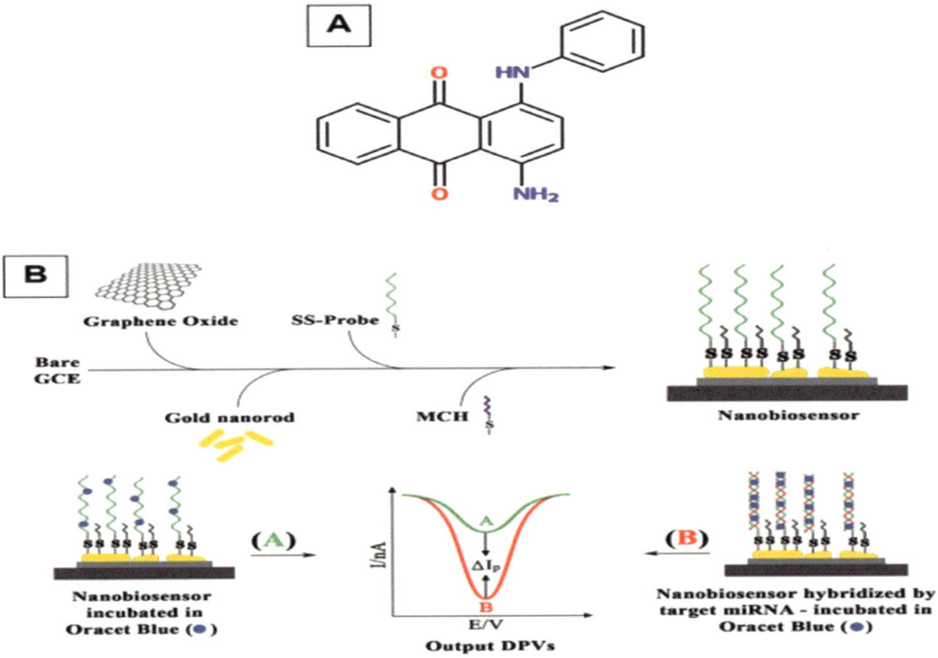

Figure 5.
(A) Molecular structure of the Oracet Blue molecule, (B) schematic illustration of the assembling and working procedure of the proposed electrochemical nanobiosensor for miR-155 detection. Reprinted with permission from Ref. [22].

voltammetry technique. This electrochemical nano biosensor method provided a linear range between 2.0 and 8.0 fM and a limit of detection of 0.6 fM. In addition, the versatility of the proposed nano biosensor has been demonstrated by direct detection of the miR-155 in plasma without the need for sample extraction and/or amplification, which can be applied in clinical applications such as early detection and/or as a predictor of drug response and prognostic trends in patients with breast cancer. Lastly, the proposed electrochemical nano biosensing method could also be used to detect any sequence of miRNAs, simply by changing the capture probe.

2.5 Cancer antigen 125 (CA125)

The cancer antigen 125 also known as mucin 16 (MUC 16) is a therapeutic tumor marker present in many ovarian cancer cells on the surface. However, some malignant diseases such as breast cancer, mesothelioma, non-Hodgkin's lymphoma can lead to increased CA125 levels. Usually, the normal blood CA125 concentration is less than 35 U mL^{-1} (units per milliliter) [22, 23]. A number of electrochemical immunosensors for detection of CA125 have been developed so far.

Recent study, Fan et al. [24] developed paper-based electrochemical immunosensor to detect cancer antigen 125 (CA125) by screen-printing method. The reduced nanocomposites of graphene oxide/thionine/gold (rGO/Thi/AuNPs) were compounded and coated for CA125 antibody (anti-CA125) immobilization and identification signal amplification on the paper electrode of the immunosensor. The principle of detection was based on the premise that the immunocomplex produced by the binding of CA125 antibody and antigen could reduce the current thionine reaction, which was corresponding to the corresponding CA125 antigen concentration. The results of the immunoassay demonstrate that the linear range of CA125 was between 0.1 U mL^{-1} and 200 U mL^{-1} with a detection limit of 0.01 U mL^{-1}.

In addition, Ravalli and coworkers [25] developed the analytical performances of a label-free impedimetric immunosensor for the detection of tumor marker CA125 using modified screen-printed graphite electrode in gold nanoparticles. In this works the immunoassay is focused on a self-assembled monolayer (SAM) of electrodeposited gold nanostructures with corresponding monoclonal antibody immobilization on screen-printed graphite electrodes as shown in **Figure 6**. The development of immunosensor each steps are characterized using cyclic voltammetry (CV) and electrochemical impedance spectroscopy (EIS) techniques. The response of the immunosensor ranged linearly ($r^2 = 0.996$) with antigen concentration varying from 0 to 100 U mL^{-1}. The approximate detection limit was found to be 6.7 U mL^{-1}. Also, the immunoassay was tested on serum samples, resulting in promising results for use with real samples.

Recently, Baradoke et al. [26] constructed Screen-printed carbon electrodes were electroplated with gold or platinum nanostructures and used as an antibody immobilization platform for CA125. In this works, nanostructured surfaces have been used as a tool for the design of immunosensors and analyses in electrochemical kinetics such as antibody immobilization, electrode surface blocking and antigen binding. In addition, the detection of CA125 was demonstrated on the coated Au and Pt nanostructured interfaces with a LOD of 419 ± 31 ng mL^{-1} and 386 ± 27 ng mL^{-1}.

2.6 Breast cancer type 1 and 2 susceptibility proteins (BRCA1 and BRCA2)

In regards to breast cancer, some hereditary gene mutations were linked with the development of cancer, mainly identified with BRCA1 and BRCA2 tumor suppressor genes [27]. In specific, the BRCA1 gene encodes a protein of 1863

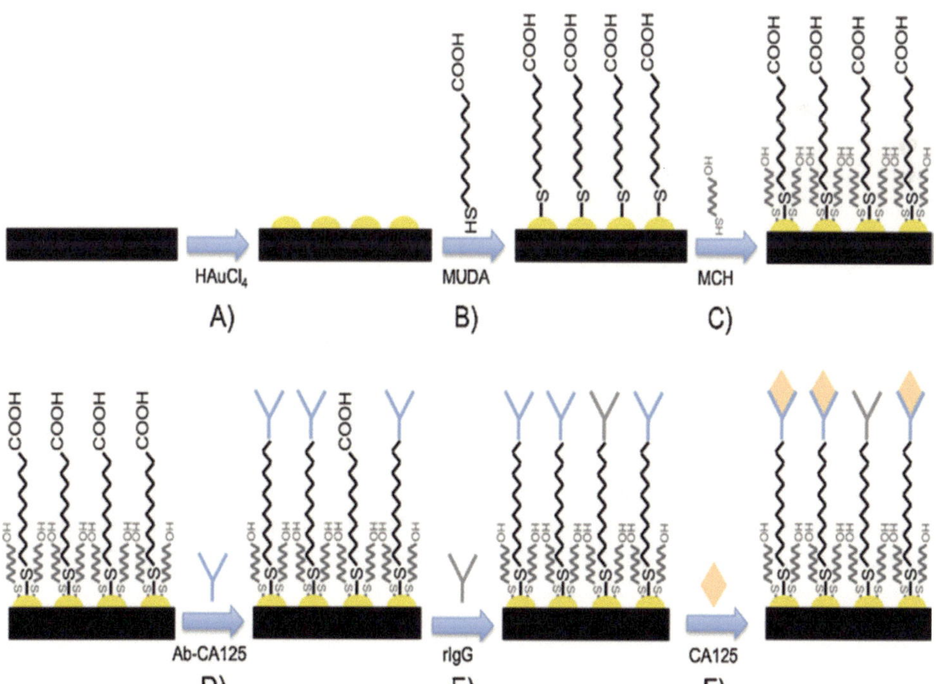

Figure 6.
Scheme of CA125 immunosensor developed: (A) electrodepositing of AuNPs on SPGE, (B) functionalization of gold nanoparticles with MUDA, (C) mixed SAM formation with MCH, (D) activation of COOH groups with EDAC/NHS and Ab-CA125 immobilization, (E) blocking step with rIgG, (F) Ab-Ag affinity reaction. Reprinted with permission from Ref. [26].

amino acid implicated in genomic stability. Mutations in this gene are characterized as closely associated with and early onset of family breast cancer syndrome. These are also responsible for controlling and managing the checkpoints and cell division of the cell cycle. Mutations in the BRCA1 and BRCA2 genes are linked to increase of breast cancer and are essential for about 21–40% of hereditary cases of breast cancer [28]. BRCA1 protein expression was reported to be decreased in 30% of sporadic cases of breast cancer [29]. The extent of the BRCA1 protein reduction depends on the extent of the breast cancer and is inversely to the expression of BRCA2 protein used as a tool for the treatment of sporadic breast cancer [30]. Furthermore, BRCA2 can be used for breast cancer as both a prognostic and a screening biomarker.

Currently, Shahrokhian and Salimian [31] developed an ultrasensitive label-free electrochemical DNA (E-DNA) sensor based on conducting polymer/reduced graphene-oxide platform has been developed for the detection of BRCA1 gene. An electrochemical technique was used as a simple and easy to control method for the electrochemical reduction of graphene oxide and also for the electropolimerization of the monomer of pyrrole 3 carboxylic acid. The signal produced from the E-DNA sensor uses CV, DPV and EIS methods to detect the redox probe's electrochemical behavior. This sensor allows BRCA1 to be quantitatively determined in the linear range of 10 fM to 0.1 μM with a low detection limit as 3 fM. In addition, the modified electrode was effectively used in blood plasma samples to accurately determine the trace amounts of the DNA target.

Furthermore, recently developed novel immunoassay based on multiple polymer signal amplification technique for BRCA1 protein recognition. The developed immunoassay processed by poly (dopamine-beta cyclodextrine-cetyl trimethylammonium bromide) doped by silver nanoparticles (P[DA-β-CD/CTAB])-AgNPs and functionalized mesoporous silica matrix (MCM-41-SO_3H) produced on the glassy carbon electrode with a large surface area that has been designed to provide a new device for the immobilization of primary antibodies and outstanding conductivity. MCM-41-SO_3H provides the appropriate volume of pores and functional groups to detect further horseradish peroxidase-labeled antibodies and improve conductivity to further amplify the electrochemical signal. The experimental immunoassay indicates adequate analytical efficiency for BRCA1 screening with a linear range of 0.01565–10 pg mL^{-1} (DPV) and 0.625–20 pg mL^{-1} (SWV) and a low quantification value of 0.003 pg mL^{-1} [32].

In another study, label free DNA biosensor on a modified magnetic bar carbon paste electrode for BRCA1 mutation detection. In this research works, firstly, Fe_3O_4-RGO nanoparticles were synthesized, accompanied by physical adsorption of the synthesized nanoparticles composite to the built magnetic bar carbon paste electrode (MBCPE) as shown in **Figure 7**. Using PANHS leads to decreasing electrode preparation, possessing an excellent selectivity for determination of BRCA1. However, the composite of the nanoparticles are linked with using 1-pyrene butyric acid-N-hydroxy succinimide ester (PANHS) as a detection of DNA sequence also (BRCA1 5382 insC mutation detection) strands were immobilized on the surface of the electrode for exact incubation time. By using EIS technique the linear range (1.0×10^{-18} mol L^{-1} -1.0×10^{-8} mol L^{-1}) and the low detection value of 2.8×10^{-19} mol L^{-1} [33].

2.7 Interleukin-6 and -8 (IL-6 and IL-8)

Interleukin-6 (IL-6) is a leukocyte-secreted 21 kDa glycoprotein, related to as both a pro- and an anti-inflammatory cytokine because it has roles in both directions. Interleukin (IL)-8 is an inflammatory chemokine contained in the subfamily C-X-clinically relevant rates of IL-6 in physiologically normal situations

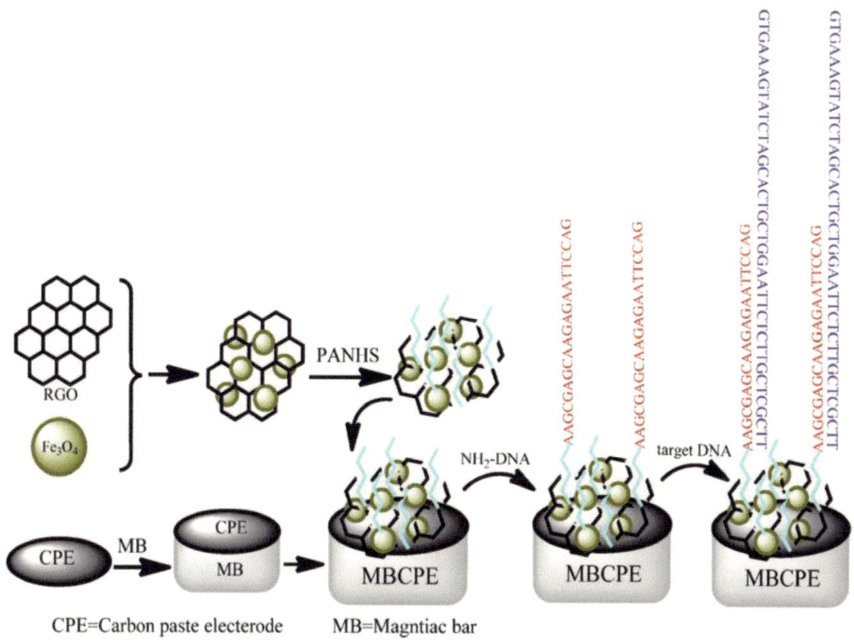

Figure 7.
Schematic representation of the modified electrochemical biosensor based on MBCPE/Fe_3O_4-RGO/PANHS platform. Reprinted with permission from Ref. [34].

have recently been recorded as 5–25 pg mL^{-1} and up to 1000 pg mL^{-1} in sepsis patients [34]. IL-8 is expressed by certain cell types, including activated monocytes and macrophages, a wide range of epithelial cells and fibroblasts [35].

For the detection of interleukin 1β in human serum and saliva, new impedimetric immunosensor was prepared using semi-conductive poly (2-thiophen-3-yl-malonic acid) (P3-TMA) as matrix material for immobilization and anti-IL-1β antibody as a component for biorecognition. P3-TMA added a lot of antibody binding in the presence of carboxyl groups. EIS and CV techniques were used d to monitor the detection of IL-1β antigen concentration in the range of 0.01–3 pg mL^{-1} with the detection value of 3 fg mL^{-1} [36].

Using a similar principle, an impedimetric immunosensor for highly sensitive detection of IL-8 is in human serum and saliva samples. 6-phosphonohexanoic acid (PHA) was used to label the anti-IL-8 antibody. In addition, anti-IL8 antibody interaction to IL8 antigen was observed using SFI (single frequency impedance) technique as shown in **Figure 8**. EIS technique was applied for the interrogation of IL-6 level in the linear range of 0.02–3 pg mL^{-1} with a detection value of 6 pg mL^{-1} and good stability (7 weeks) [37].

2.8 Vascular endothelial growth factor (VEGF)

Specific tyrosine kinase receptors divided into subtypes of VEGFR-1, VEGFR-2, and VEGFR-3 are vascular endothelial growth factor receptors (VEGFRs). VEGFR-2 activates most angiogenic processes among the three forms. VEGFR-2 is an important biomarker for breast cancer with a blood level of >15 ng L^{-1} indicating the presence of tumors of the breast cancer [38].

For detection of VEGFR-2, a sandwich immunoassay was designed to detect VEGFR-2, by immobilizing anti-VEGFR-2 Ab1using chitosan/rGO/thionin-modified GCE as shown in **Figure 9**. An HRP-labeled Ab2 was used to identify antibody,

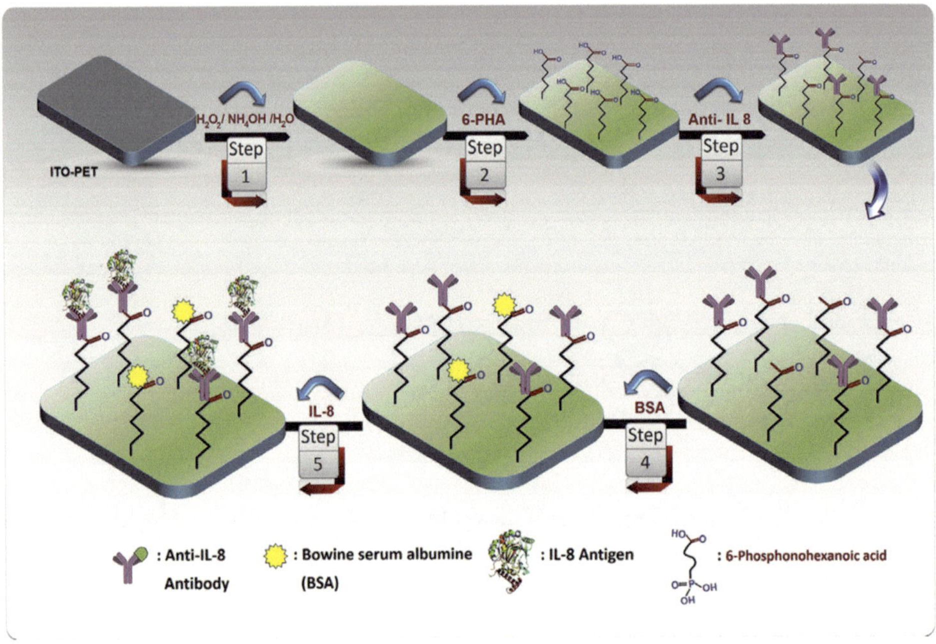

Figure 8.
Schematic representation of the impedimetric immunosensor IL-8 detection. Reprinted with permission from Ref. [38].

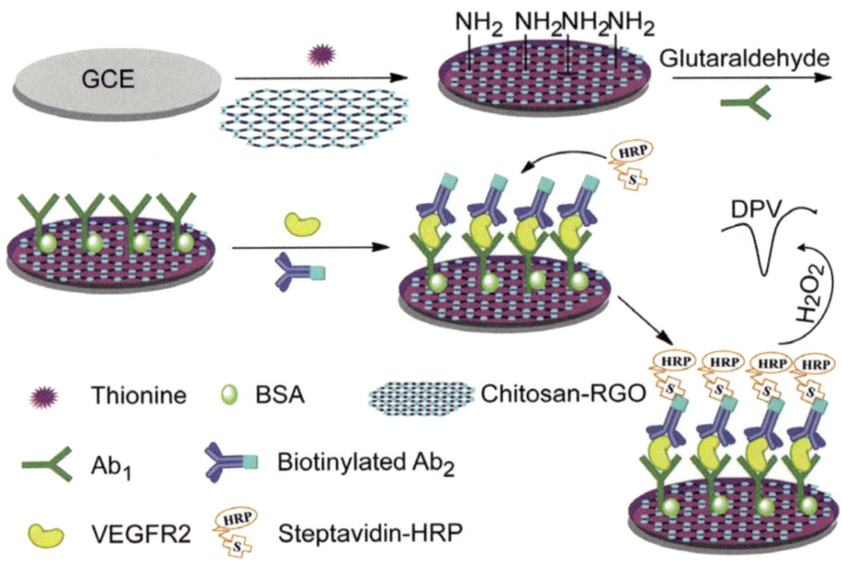

Figure 9.
Schematic representation of the electrochemical biosensing for VEGFR2 detection. Reprinted with permission from Ref. [40].

catalyzing thionine oxidation by H_2O_2. VEGFR-2 was quantified by DPV with a detection limit of 0.28 pM in the linear concentration range of 0.4–86.0 pM [39]. For its detection, VEGFR, sensitive label-free impedimetric sensor fabricated on molecular impressed polymer (MIP) as a biomimetic receptor coupled with screen-printed electrodes. Next, o-phenylenediamine (oPD) electropolymerization was conducted on graphite-screen-printed electrodes in the presence of VEGF molecules through cyclic voltammetry. The single-use MIP-based sensor demonstrated good analytical efficiency for VEGF detection from 20 to 200 pg mL^{-1} with a detection limit of 0.08 pg mL^{-1} using with EIS technique [40].

2.9 Cluster differentiation 146 Ag (CD-146)

Cluster differentiation 146 Ag (CD-146) is a molecule of cell adhesion that belongs to the superfamily of immunoglobulins. It is identified as a progression marker for melanoma (melanoma adhesion molecule antigen) and breast cancer. The normal level of CD-146 in blood serum of healthy individuals is generally 309 μg L^{-1} [41].

For the identification of CD-146, a sandwich-based amperometric immunosensor was manufactured in which rGO-tetra ethylene pentaamine (TEPA) enhanced GCE antibody (Ab1) was immobilized as shown **Figure 10**. This improvement offered the electrode a large number of amino groups to improve the loading potential of antibodies. The secondary Ab was controlled with colloidal sphere TiO_2 and nanoparticles Au/Pd and assay was conducted by calculating the amperometric reaction to electrocatalytic reduction of H_2O_2. However, the immunosensor displayed a wide linear range in 0.0050–20 ng mL^{-1}, a low limit detection value of 1.6 pg mL^{-1} [42].

Furthermore, biomimetic mussel-inspired polydopamine coating photoelectrochemical biosensing chip was constructed to detect CD146. The CdS/TiO_2-ITO chip was designed using the electrodeposition process to deposit CdS on the TiO_2-ITO chip. In addition, the PDA (polydopamine), developed by DA (dopamine) self-polymerization, was anchored on the CdS/TiO_2-ITO (cadmium sulphide/titanium dioxide-indium tinoxide) chip surface through its strong adhesivity and specific interactions such as electrostatic attractions or covalent bindings. Also, without using external crosslinkers, PDA/CdS/TiO_2-ITO chips could be used for direct immobilization of antibodies. By measuring the photocurrent responses to different concentrations of CD146, quantitative determination of CD146 was achieved based on this principle. The photocurrent decreased linearly from 1 pg mL^{-1} to 20 ng mL^{-1} with an increase in CD146 concentration and a detection value of 0.3 pg mL^{-1} [43].

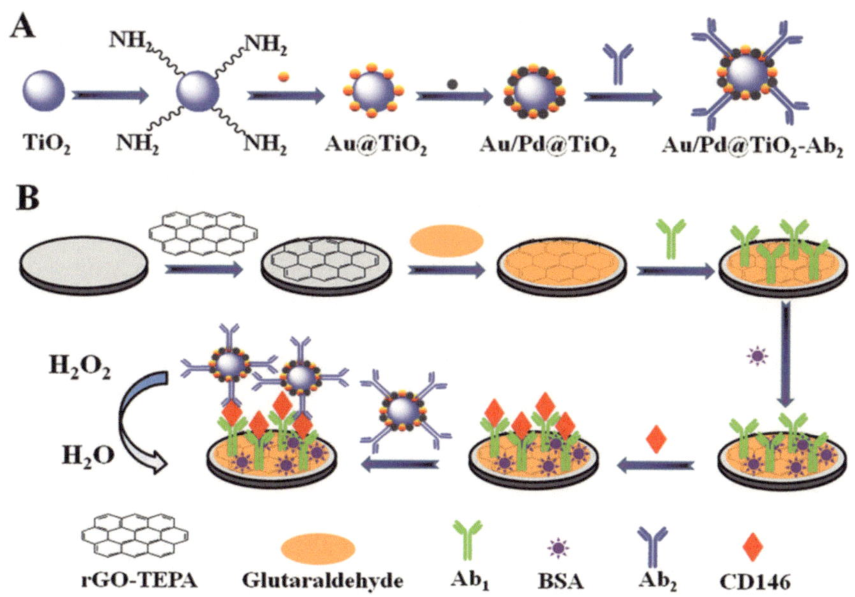

Figure 10.
Schematic representation of the preparation of Au/Pd@TiO_2-Ab2 (A) and immunosensor (B). Reprinted with permission from Ref. [43].

2.10 Other important biomarkers

Biomarkers such as tenascin-C (TN-C); p53 tumor suppressor protein (p53); DNA methyltransferase (DNA-MTase); estrogen receptor (ER); matrix metallopeptidase 9 (MMP-9); apurinic/apyrimidinic endonuclease 1 (APE1); mucin-like carcinoma-associated antigen (MCA); murine double minute 2 (MDM2); nuclear factor kappa B (NF-κB) are other important biomolecules that are known as tumor markers for breast cancer [44].

Tenascin-C is a large extracellular matrix protein newly expressed during tissue remodeling processes including angiogenesis, inflammation, and tumor growth. Tenascin-C is especially overexpressed in tumor tissues [45]. For the detection of TN-C a chemiluminescence (CL) based voltammetric immunosensor was prepared using carboxylated carbon nanoparticles (cCNPs) and conjugated with magnetic beads (MBs). In addition, an aptamer labeled with the CL reagent N-(4-aminobutyl)-N-ethylisoluminol was adsorbed to form labeled aptamer modified cCNPs-MBs on the surface of carboxy-modified magnetic carbon nanoparticles. When the tenascin-C sample is applied, it will associate with the labeled aptamer to build a complex with the labeled aptamer. This complex is measured by CL whose amplitude is linearly linked to the tenascin C concentration by 1 pM to 1 nM and the detection value as 0.4 pM respectively [46].

Tumor suppressor protein (p53) is an important transcription factor in regulating cellular responses to stress factors. Loss of p53 activity induces tumor activation and gene mutation, resulting in a conformational alteration in the p53 protein structure [47]. The level of p53 protein in serum samples from cancer patients varies from 0.52 ± 0.23 ng mL^{-1} to 1.03 ± 0.59 ng mL^{-1} [48]. Another, nanocomposite based ultrasensitive electrochemical immunosensor for p53 was designed using polycysteine/graphene quantum dots/gold nanoparticle. In this study, p53-antibody was immobilized on a green and biocompatible nanocomposite comprising L-cysteine (P-Cys) as a conductive matrix and gold nanoparticles (GNPs) as a dual amplification component of graphene quantum dots (GQDs). This approach facilitated the linear analysis of p53 in the range from 0.0244 to 0.369 pM (SWV technique) and 0.195–50 pM (DPV technique) with a lower limit of 12.1 fM [49].

DNA methyltransferase (DNAMTase) is a widespread epigenetic alteration in both prokaryotes and eukaryotes and plays a crucial role in controlling gene expression, genomic stability and cell growth. Most tumors such as thyroid, liver, heart, prostate and breast cancer have over-expression of DNAMTase [50]. Further, DNAMTase was detected by label-free electrochemical biosensor for identification and inhibitor screening of methyltransferase activity based on graphene quantity and enzyme-catalyzed reaction. However, in this work, the modified HRP catalyzed the 3, 3′, 5, 5′-tetramethylbenzidine hydrogen peroxide-mediated oxidation resulting in an electrochemical signal production. The proposed biosensor achieved sensitivity detection of MTase activity within the range of 1–40 U mL^{-1} with a detection limit of 0.3 U mL^{-1} [51].

The biomarker of the estrogen receptor alpha (ERα) is a receptor and transcription factor with nuclear hormones. It controls gene expression and inhibits the proliferation and differentiation of cells in the target tissue. The occurrence of elevated ERα rates in the breast epithelium that implies an increased risk of breast cancer, indicating both the function of ERα in initiation and cancer progression [52]. For ERα detection, a electrochemical magneto immunosensing platform was developed taking an anti-human ERα antibodies modified SPCE. The antibody acquisition was immobilized on modified magnetic beads in carboxylic acid compounds, while the biotinylated antibody was labeled with a streptavidin-HRP conjugate. In addition, determining the target ERαprotein with a detection limit of 19 pg mL^{-1} evaluated the applicability of the integrated disposable magneto immunosensor [53].

Matrix metalloproteinase 9 (MMP-9) is an extracellular 92 kDa protease belonging to a family of endopeptidases dependent on zinc and calcium. There was an improvement in expression of MMP-9 in a number of different tumors relative to healthy subjects, with an overall positive association between tumor aggressiveness and activity levels of MMP-9 [54]. For detection of MMP-9, SERS (surface-enhanced Raman scattering) nano-tags integrated magnetic-separation biosensor was designed with whole blood. The silica coated Ag SERS nano-tags built as labels were used to identify MMP-9 in unprocessed blood samples in a quick and accurate MSB immune sensor. The results showed the sensitive and reproducible response constructed in whole blood to the concentration of MMP-9 in the range up to 100 ng mL^{-1} with a detection limit of 1 pg mL^{-1} [55].

Human apurinic/apyrimidinic endonuclease 1 (APE1) is an intracellular multifunctional enzyme, also known as the redox impact factor 1. For detection of APE1, a novel electrochemical biosystem was designed by immobilizing antibody-APE1 modified on gold electrode. The technique skillfully incorporates immunoassay through an intricate template of enzyme activity study. The APE1 biosensor detection limit is as low as 0.00518 U mL^{-1}. The technique will monitor enzyme activity controlled by an APE1 inhibitor and its isozyme discrimination [56].

Mucin 1 (MUC1) protein is a membrane-associated glycoprotein containing 31 amino acids in the hydrophobic domain, 69 amino acids in the cytoplasmic domain and 20 identical amino acids per repeat in the extracellular domain. MUC1 is also a well-known tumor marker present in a range of malignant tumors [57]. For detection of MUC1, carboxylic group of disposable electrochemical immunosensors rich in graphene oxide for the identification of biomarkers with methylene blue using with human serum samples. The authors investigated highly conductive surfaces of carboxylic group rich graphene oxide on screen-printed carbon electrodes in this process. The established immunosensor demonstrated good detection range (0.1–2 U mL^{-1}), for MUC1 with outstanding linearity with a detection limit of 0.04 U mL^{-1} by using the differential pulse voltammetry (DPV) technique [58].

3. Future prospects and challenges

In recent years, the development of biosensors for biomarkers for breast cancer has received a lot of attention. However, the developments of biomarkers and the innovation of diagnostic tools for early detection of breast cancer are still in their early stages. While electrochemical immunoassays were very successful bio transducers, biomarkers established for breast cancer are needed to test their specificity, responsiveness and efficiency against the diagnostic standards created. The production and progression of these advanced cancer screening systems will aid in the early stages of accelerated clinical cancer diagnoses. Nonetheless, proposed detection approaches for biomarker detection of cancer necessarily require standardization of pre- and post-analytical protocols such as sample preparation, storage and optimization of experimental conditions for true validity of assays and more genuine output of the biosensor produced. Although very low LODs have been obtained by electrochemical biosensors, they typically convey multi-step mark strategies that complicate the experimental activity. For future works, the development another problem is that owing to its low accuracy and reliability, few portable electrochemical instruments are in clinical usage. Therefore, robust biosensor-based POC devices are required of ultrasensitive electrochemical label-free methods will be a great potential. Researchers must train the electrochemical biosensor to solve their reliability problems with a significant number of clinical samples. The development of wireless micro/nano electrochemical biosensors is an ideal option for in vivo detection, as they can work in an invasively style. The approachable properties of electrochemical instruments improve the performance of cancer diagnostics and therapy monitoring. With further advancement and funding, these handheld instruments are anticipated to improve cancer diagnosis, rendering diagnostic findings accessible in a matter of minutes at the patient bedside or practitioner's office.

4. Conclusion

Novel electrochemical techniques for the detection of biomarkers of breast cancer have been established using the remarkable progress in nanotechnology and biosensor techniques. Effective electrochemical detection systems are combined with immunology guidelines to host different antibody-complex reactions on the transducer surface and quantitatively identify the biomarkers. Current electrochemical biosensors, such as DNA or immunosensors, have extraordinary sensitivity that is important for early detection of cancer. The integration of electrochemical instruments with nano-scale materials provides a unique multiplexing mechanism for various cancer marker simultaneous measurements. However, before marketing the developed biosensors for actual clinical practices for the detection of biomarkers for breast cancer, important gaps and approaches need to be addressed and implemented.

Acknowledgements

All the authors of the manuscript thank and acknowledge support of their respective Universities and Institutes.

Conflict of interest

The authors declare no conflicts of interest to disclose in relation to this book chapter.

Author details

Stephen Rathinaraj Benjamin[1*] and Fabio de Lima[2]

1 Department of Biomedical Engineering, Federal University of Pernambuco, Recife, PE, Brazil

2 Department of Computer Science, Federal University of Mato Grosso do Sul, Campo Grande, MS, Brazil

*Address all correspondence to: biotechstephen@gmail.com

References

[1] WHO. WHO Cancer Fact Sheet 2018. 2018. Available from: https://www.who.int/news-room/fact-sheets/detail/cancer

[2] WHO. WHO Breast cancer. WHO. 2017. Available from: https://www.who.int/cancer/prevention/diagnosis-screening/breast-cancer/en/

[3] Wei PL, Lee LT, Tseng LM, Huang KW. Validation of assaying carcinoembryonic antigen in human serum by using immunomagnetic reduction. Scientific Reports. 2018;**8**(1):10002

[4] Rizwan M, Elma S, Lim SA, Ahmed MU. AuNPs/CNOs/SWCNTs/chitosan-nanocomposite modified electrochemical sensor for the label-free detection of carcinoembryonic antigen. Biosensors and Bioelectronics. 2018;**107**:211-217

[5] Gao Y-S, Zhu X-F, Xu J-K, Lu L-M, Wang W-M, Yang T-T, et al. Label-free electrochemical immunosensor based on Nile blue A-reduced graphene oxide nanocomposites for carcinoembryonic antigen detection. Analytical Biochemistry. 2016;**500**:80-87

[6] Lu W, Cao X, Tao L, Ge J, Dong J, Qian W. A novel label-free amperometric immunosensor for carcinoembryonic antigen based on Ag nanoparticle decorated infinite coordination polymer fibres. Biosensors and Bioelectronics. 2014;**57**:219-225

[7] Li H, Qin J, Li M, Li C, Xu S, Qian L, et al. Gold-nanoparticle-decorated boron-doped graphene/BDD electrode for tumor marker sensor. Sensors and Actuators, B: Chemical. 2020;**302**:127209

[8] DeSantis C, Naishadham D, Jemal A. Cancer statistics for African Americans, 2013. CA: A Cancer Journal for Clinicians. 2013;**63**(3):151-166

[9] De Oliveira RAG, Materon EM, Melendez ME, Carvalho AL, Faria RC. Disposable microfluidic immunoarray device for sensitive breast cancer biomarker detection. ACS Applied Materials and Interfaces. 2017;**9**(33):27433-27440

[10] Qin D, Jiang X, Mo G, Feng J, Yu C, Deng B. A novel carbon quantum dots signal amplification strategy coupled with sandwich electrochemiluminescence immunosensor for the detection of ca15-3 in human serum. ACS Sensors. 2019;**4**(2):504-512

[11] Wang A-J, Zhu X-Y, Chen Y, Luo X, Xue Y, Feng J-J. Ultrasensitive label-free electrochemical immunoassay of carbohydrate antigen 15-3 using dendritic Au@Pt nanocrystals/ferrocene-grafted-chitosan for efficient signal amplification. Sensors and Actuators, B: Chemical. 2019;**292**:164-170

[12] Khoshroo A, Mazloum-Ardakani M, Forat-Yazdi M. Enhanced performance of label-free electrochemical immunosensor for carbohydrate antigen 15-3 based on catalytic activity of cobalt sulfide/graphene nanocomposite. Sensors and Actuators, B: Chemical. 2018;**255**:580-587

[13] WHO. Press announcements—FDA approves new treatment for certain advanced or metastatic breast cancers. FDA News Release. 2017

[14] Olsen D, Trøst Jørgensen J. Companion diagnostics for targeted cancer drugs—Clinical and regulatory aspects. Frontiers in Oncology. 2014;**4**(105):1-8

[15] Sharma S, Zapatero-Rodríguez J, Saxena R, O'Kennedy R, Srivastava S. Ultrasensitive direct impedimetric immunosensor for detection of serum HER2. Biosensors and Bioelectronics. 2018;**106**:78-85

[16] Shamsipur M, Emami M, Farzin L, Saber R. A sandwich-type electrochemical immunosensor based on in situ silver deposition for determination of serum level of HER2 in breast cancer patients. Biosensors and Bioelectronics. 2018;**103**:54-61

[17] Lah ZMANH, Ahmad SAA, Zaini MS, Kamarudin MA. An electrochemical sandwich immunosensor for the detection of HER2 using antibody-conjugated PbS quantum dot as a label. Journal of Pharmaceutical and Biomedical Analysis. 2019;**174**:608-617

[18] Freitas M, Neves MMPS, Nouws HPA, Delerue-Matos C. Quantum dots as nanolabels for breast cancer biomarker HER2-ECD analysis in human serum. Talanta. 2020;**208**:120430

[19] Wang H, Sun J, Lu L, Yang X, Xia J, Zhang F, et al. Competitive electrochemical aptasensor based on a cDNA-ferrocene/MXene probe for detection of breast cancer marker Mucin1. Analytica Chimica Acta. 2020;**1094**:18-25

[20] Tian L, Qian K, Qi J, Liu Q, Yao C, Song W, et al. Gold nanoparticles superlattices assembly for electrochemical biosensor detection of microRNA-21. Biosensors and Bioelectronics. 2018;**99**:564-570

[21] Azimzadeh M, Rahaie M, Nasirizadeh N, Ashtari K, Naderi-Manesh H. An electrochemical nanobiosensor for plasma miRNA-155, based on graphene oxide and gold nanorod, for early detection of breast cancer. Biosensors and Bioelectronics. 2016;**77**:99-106

[22] Gupta D, Lis CG. Role of CA125 in predicting ovarian cancer survival—A review of the epidemiological literature. Journal of Ovarian Research. 2009;**2**(1):2-13

[23] Ghosh I, Bhattacharjee D, Das AK, Chakrabarti G, Dasgupta A, Dey SK. Diagnostic role of tumour markers CEA, CA15-3, CA19-9 and CA125 in lung cancer. Indian Journal of Clinical Biochemistry. 2013;**28**(1):24-29

[24] Fan Y, Shi S, Ma J, Guo Y. A paper-based electrochemical immunosensor with reduced graphene oxide/thionine/gold nanoparticles nanocomposites modification for the detection of cancer antigen 125. Biosensors and Bioelectronics. 2019;**135**:1-7

[25] Ravalli A, Pilon Dos Santos G, Ferroni M, Faglia G, Yamanaka H, Marrazza G. New label free CA125 detection based on gold nanostructured screen-printed electrode. Sensors and Actuators, B: Chemical. 2013;**179**(31):194-200

[26] Baradoke A, Jose B, Pauliukaite R, Forster RJ. Properties of Anti-CA125 antibody layers on screen-printed carbon electrodes modified by gold and platinum nanostructures. Electrochimica Acta. 2019;**306**:299-306

[27] Godet IM, Gilkes D. BRCA1 and BRCA2 mutations and treatment strategies for breast cancer. Integrative Cancer Science and Therapeutics. 2017;**4**(1):1-17

[28] King M-C. Breast and ovarian cancer risks due to inherited mutations in BRCA1 and BRCA2. Science. 2003;**302**(5645):643-646

[29] Yang Q, Sakurai T, Mori I, Yoshimura G, Nakamura M, Nakamura Y, et al. Prognostic significance of BRCA1 expression in Japanese sporadic breast carcinomas. Cancer. 2001;**92**(1):54-60

[30] Hedau S, Batra M, Singh U, Bharti A, Ray A, Das B. Expression of BRCA1 and BRCA2 proteins and their correlation with clinical staging in breast cancer. Journal of

Cancer Research and Therapeutics. 2015;**11**(1):158-163

[31] Shahrokhian S, Salimian R. Ultrasensitive detection of cancer biomarkers using conducting polymer/electrochemically reduced graphene oxide-based biosensor: Application toward BRCA1 sensing. Sensors and Actuators B: Chemical. 2018;**266**:160-169

[32] Hasanzadeh M, Feyziazar M, Solhi E, Mokhtarzadeh A, Soleymani J, Shadjou N, et al. Ultrasensitive immunoassay of breast cancer type 1 susceptibility protein (BRCA1) using poly (dopamine-beta cyclodextrine-Cetyl trimethylammonium bromide) doped with silver nanoparticles: A new platform in early stage diagnosis of breast cancer and effici. Microchemical Journal. 2019;**145**:778-783

[33] Jahanbani S, Benvidi A. A novel electrochemical DNA biosensor based on a modified magnetic bar carbon paste electrode with Fe_3O_4NPs-reduced graphene oxide/PANHS nanocomposite. Materials Science and Engineering: C. 2016;**68**:1-8

[34] Molano Franco D, Arevalo-Rodriguez I, Roqué i Figuls M, Zamora J. Interleukin-6 for diagnosis of sepsis in critically ill adult patients. Cochrane Database of Systematic Reviews. 2015;7:1-16

[35] Mukaida N, Harada A, Yasumoto K, Matsushima K. Properties of pro-inflammatory cell type-specific leukocyte chemotactic cytokines, interleukin 8 (IL-8) and monocyte chemotactic and activating factor (MCAF). Microbiology and Immunology. 1992;**36**(8):773-789

[36] Aydın EB, Aydın M, Sezgintürk MK. Highly sensitive electrochemical immunosensor based on polythiophene polymer with densely populated carboxyl groups as immobilization matrix for detection of interleukin 1β in human serum and saliva. Sensors and Actuators B: Chemical. 2018;**270**:18-27

[37] Aydın EB, Sezgintürk MK. An impedimetric immunosensor for highly sensitive detection of IL-8 in human serum and saliva samples: A new surface modification method by 6-phosphonohexanoic acid for biosensing applications. Analytical Biochemistry. 2018;**554**:44-52

[38] Reeves KW, Ness RB, Stone RA, Weissfeld JL, Vogel VG, Powers RW, et al. Vascular endothelial growth factor and breast cancer risk. Cancer Causes & Control. 2009;**20**(3):375-386

[39] Wei T, Tu W, Zhao B, Lan Y, Bao J, Dai Z. Electrochemical monitoring of an important biomarker and target protein: VEGFR2 in cell lysates. Scientific Reports. 2015;**4**(1):3982. 1-7

[40] Bozal-Palabiyik B, Lettieri M, Uslu B, Marrazza G. Electrochemical detection of vascular endothelial growth factor by molecularly imprinted polymer. Electroanalysis. 2019;**31**(8):1458-1464

[41] Zhang B, Li L, Feng L, Zhang Y, Zeng X, Feng J, et al. Elevated levels of soluble and neutrophil CD146 in active systemic vasculitis. Laboratory Medicine. 2009;**40**(6):351-356

[42] Wang Y, Li Y, Ma H, Guo A, Du B, Yan T, et al. An ultrasensitive electrochemical immunosensor for the detection of CD146 based on TiO_2 colloidal sphere laden Au/Pd nanoparticles. The Analyst. 2015;**140**(10):3557-3564

[43] Ma H, Yan T, Zhang Y, Gao P, Pang X, Du B, et al. A biomimetic mussel-inspired photoelectrochemical biosensing chip for the sensitive detection of CD146. The Analyst. 2015;**140**(15):5019-5022

[44] Labib M, Sargent EH, Kelley SO. Electrochemical methods for the analysis of clinically relevant biomolecules. Chemical Reviews. 2016;**116**(16):9001-9090

[45] Kang WJ, Chae JR, Cho YL, Lee JD, Kim S. Multiplex imaging of single tumor cells using quantum-dot-conjugated aptamers. Small. 2009;**5**(22):2519-2522

[46] Wu Y, Nie F, Xia D. Chemiluminescence assay for the glycoprotein tenascin-C based on aptamer-modified carboxylated magnetic carbon nanoparticles. Microchimica Acta. 2015;**182**(1-2):227-232

[47] Chen A, Bao Y, Ge X, Shin Y, Du D, Lin Y. Magnetic particle-based immunoassay of phosphorylated p53 using protein cage templated lead phosphate and carbon nanospheres for signal amplification. RSC Advances. 2012;**2**(29):11029-11034

[48] Attallah AM, Abdel-Aziz MM, El-Sayed AM, Tabll AA. Detection of serum p53 protein in patients with different gastrointestinal cancers. Cancer Detection and Prevention. 2003;**27**(2):127-131

[49] Hasanzadeh M, Baghban HN, Shadjou N, Mokhtarzadeh A. Ultrasensitive electrochemical immunosensing of tumor suppressor protein p53 in unprocessed human plasma and cell lysates using a novel nanocomposite based on poly-cysteine/graphene quantum dots/gold nanoparticle. International Journal of Biological Macromolecules. 2018;**107**:1348-1363

[50] Miyamoto K, Fukutomi T, Akashi-Tanaka S, Hasegawa T, Asahara T, Sugimura T, et al. Identification of 20 genes aberrantly methylated in human breast cancers. International Journal of Cancer. 2005;**116**(3):407-414

[51] Peng X, Hu T, Bao T, Zhao L, Zeng X, Wen W, et al. A label-free electrochemical biosensor for methyltransferase activity detection and inhibitor screening based on graphene quantum dot and enzyme-catalyzed reaction. Journal of Electroanalytical Chemistry. 2017;**799**:327-332

[52] Ali S, Coombes RC. Estrogen receptor alpha in human breast cancer:occurrence and significance. Journal of Mammary Gland Biology and Neoplasia. 2000;**5**(3):271-281

[53] Eletxigerra U, Martinez-Perdiguero J, Merino S, Barderas R, Ruiz-Valdepeñas Montiel V, Villalonga R, et al. Estrogen receptor α determination in serum, cell lysates and breast cancer cells using an amperometric magnetoimmunosensing platform. Sensing and Bio-Sensing Research. 2016;**7**:71-76

[54] Kessenbrock K, Plaks V, Werb Z. Matrix metalloproteinases: Regulators of the tumor microenvironment. Cell. 2010;**141**(1):52-67

[55] Zhao P, Li H, Li D, Hou Y, Mao L, Yang M, et al. A SERS nano-tag-based magnetic-separation strategy for highly sensitive immunoassay in unprocessed whole blood. Talanta. 2019;**198**:527-533

[56] Zhou M, Feng C, Mao D, Yang S, Ren L, Chen G, et al. An electrochemical biosensor integrating immunoassay and enzyme activity analysis for accurate detection of active human apurinic/apyrimidinic endonuclease 1. Biosensors and Bioelectronics. 2019;**142**:111558

[57] Hu R, Wen W, Wang Q, Xiong H, Zhang X, Gu H, et al. Novel electrochemical aptamer biosensor based on an enzyme–gold nanoparticle dual label for the ultrasensitive detection of epithelial tumour marker MUC1. Biosensors and Bioelectronics. 2014;**53**:384-389

[58] Rauf S, Mishra GK, Azhar J, Mishra RK, Goud KY, Nawaz MAH, et al. Carboxylic group riched graphene oxide based disposable electrochemical immunosensor for cancer biomarker detection. Analytical Biochemistry. 2018;**545**:13-19